ELECTRONIC LEARNING

Other Titles of Interest

BP74   Electronic Music Projects
BP174   More Advanced Electronic Music Projects
BP266   Electronic Modules and Systems for Beginners

ELECTRONIC MUSIC
LEARNING PROJECTS

by

R. BEBBINGTON, M.I.S.T.C.

**BERNARD BABANI (publishing) LTD
THE GRAMPIANS
SHEPHERDS BUSH ROAD
LONDON W6 7NF
ENGLAND**

Please Note

Although every care has been taken with the production of this book to ensure that any projects, designs, modifications and/or programs, etc., contained herewith, operate in a correct and safe manner and also that any components specified are normally available in Great Britain, the Publishers do not accept responsibility in any way for the failure, including fault in design, of any project, design, modification or program to work correctly, or to cause damage to any other equipment that it may be connected to or used in conjunction with, or in respect of any other damage or injury that may be so caused, nor do the Publishers accept responsibility in any way for the failure to obtained specified components.

Notice is also given that if equipment that is still under warranty is modified in any way or used or connected with home-built equipment then that warranty may be void.

© 1993 BERNARD BABANI (publishing) LTD

First Published — May 1993

British Library Cataloguing in Publication Data
Bebbington, R.
 Electronic Music Learning Projects
 I. Title
 786.7

ISBN 0 85934 329 4

Printed and Bound in Great Britain by Cox & Wyman Ltd, Reading

Introduction

Electronics hobbyists with little or no knowledge of music often overlook any projects that have the slightest hint of musical flavour. Conversely, there are musicians who hesitate to touch electronic equipment with the end of their bow, baton or trombone slide. But music and electronics have much in common. They are made of the same stuff — vibrations!

These projects, written up in an attempt to unite both camps, feature both the musical and the electronics angles with a bit of general information sandwiched in between.

Whether you are interested in music, electronics, or both, these unusual learning projects will provide hours of entertainment.

Elementary music/electronics theory is introduced little by little, on a need-to-know basis, as you explore more practical aspects. While the primary object of this book is to help you construct some novel projects, it should also give you a better understanding of the related basic principles and techniques as you build. All the projects have been designed to cover a wide range of interests allied to music and electronics. Hopefully, you will find each chapter a useful step towards mastering both of these fascinating subjects.

Generally, the circuits are not critical as regards component layout, and instructions are sufficiently detailed to enable many of the projects to be built by beginners and younger students.

Footnote: to anxious parents
You may be relieved to learn that nothing in this book is powered by anything more dangerous than a 9-volt battery.

Roy Bebbington

Contents

Page

Chapter 1: GETTING STARTED 1
  1   Tools 1
  2   Soldering 2
  3   Printed circuit boards 2
  4   Wiring 3
  5   Mechanical construction 4

Chapter 2: BACK TO BASICS 5
  1   Electronically speaking 5
  2   Components 7
  3   Musically speaking 13
  4   The sound in music 14
  5   Octaves and scales 14
  6   The piano 14

Chapter 3: RHYTHM CHECKER 19
  1   Rhythm checker bar frame 19
  2   Rhythm checker indicator circuits 21
  3   Rhythm checker touchpad 27
  4   Finger-tapping rhythmboard 27
  5   Coda 30

Chapter 4: ELECTRONIC SOL-FA 31
  1   Mechanical construction 31
  2   The circuit 35
  3   Tuning 39

Chapter 5: DYNAMICS RANGER 41
  1   Circuit 42
  2   Construction 45

Chapter 6: TUNE-UP BOX 49
  1   Construction 49
  2   The circuit 51
  3   Variations 53

		Page
Chapter 7: MELODY RANGER	57
1	Vibrato	57
2	Mechanical construction	59
3	Circuit	59
4	Vibrato oscillator	62
5	Circuit construction	63

Chapter 8: APPEALING HANDBELLS 67
1 Mechanical construction 67
2 The circuit 69

Chapter 9: THE ELEXYLOPHONE 73
1 The circuit modules 73
2 Construction 77
3 Variations 80

Chapter 10: THE RHYTHM SETTER 83
1 Introduction 83
2 Theme 83
3 And variations 85
4 First movement 85
5 Second movement 87
6 Coda 92
7 Calibration 92

Chapter 11: THE GLIDAPHONE 95
1 Mechanical construction 95
2 The circuit 98
3 Shake it, slide it! 101

Chapter 12: THE CHORDMAKER 103
1 The circuit 105
2 Construction and interconnections 107

List of Figures 113

Chapter 1

GETTING STARTED

If you are a newcomer to electronics, it is a good plan to browse through the projects in this book before taking up the soldering iron. Also try to make up one of the easier circuits before attempting the more ambitious projects. Success on one of the simpler circuits will give you a little experience and boost your confidence to build something more complex. Most circuits are grouped into stages, so it is helpful to check each stage as you wire it before proceeding with the next. Always check the wiring before connecting the battery.

As you progress, you will realise that some of the circuits can be adapted and used for other projects. The different types of oscillators used for tone generators can often be interchanged, or other amplifiers fitted. In most cases it is best to make up the circuit first and then suitably house it when you are happy with it. As you will see, some of the components need to be selected on test, for tuning or to balance the outputs. These components are best mounted on terminal posts to make them easily accessible. This will prevent those untidy *bird's-nests* (we all know them!) that occur when circuits have to be disturbed to change components. You will find that one of the monthly magazines such as *Everyday with Practical Electronics* often features hints on constructional articles for beginners, and the advert pages provide a useful source of components.

Don't be afraid to improvise and adapt. You will learn by experience, but do make a note of failures as well as successes. You need to know what doesn't work as well as what does!

1 Tools
Some basic requirements are:

- a small pair of side cutters for cutting and stripping wires;
- a small pair of pliers for gripping components, bending the wire ends, and so on;
- a small soldering iron for connecting the components of a circuit together — a soldering iron rated at 15 watts or

25 watts has sufficient power for general electronics;
a reel of resin-cored solder;
a junior hacksaw for cutting wood and metal;
a hand-drill and bits for front-panel holes;
a flat file for shaping and smoothing.

A small multimeter for measuring amps, volts and ohms is a very useful accessory, but not essential.

2 Soldering

Soldering can be a little tricky at first, but here are a few hot tips to help you make a successful joint:

Make sure the ends of wires to be joined are clean.
Secure the connections mechanically if possible.
If necessary, clean off the tip of the hot iron with a file and 'tin' it by applying a little solder.
Apply the tip of the hot iron to the joint and melt a little solder on the joint itself. This ensures that the joint is hot enough to accept the solder, and avoids what we call 'dry' joints.
Don't apply excessive heat to transistors. Solder them quickly, or grip each wire leg at the knee with a pair of pliers to shunt the heat away while soldering the toe.

You may also find useful book number BP324 *The Art of Soldering* by the same publisher as this book.

3 Printed Circuit Boards

The printed circuit boards (pcbs) used for these projects are shown as stripboards, but experienced constructors may like to make their own pcbs from copper-clad panels if they have the necessary skill. Personally, my only problem with 0.1 in stripboard is that it is not easy on the eyes.

A little too much solder on adjacent pins of an IC, or a whisker of copper, and you have a short-circuit between the tracks that is difficult to see. Recently, I came up with an idea that virtually converts 0.1 in into more manageable 0.2 in track. When mounting an IC on the pcb, alternate pins of the IC were splayed out to lie along the top surface of the pcb instead of pushing them through the holes. The rest of the pins are taken through the stripboard holes as normal and soldered on the trackside, with the luxury of an extra track either side. The pins on the top surface are 0.2 in apart which leaves plenty of room to connect wires or components, with much less eye-strain. The spare tracks can be used to link components, or pins that lie on opposite sides of the IC.

Although not shown, the ICs can be fitted in holders if desired. This makes it easy to change ICs or to re-use them for another project — and the pins don't get overheated or damaged in the process! If holders are used, then the ICs should be fitted last after the wiring has been checked. CMOS ICs should be handled carefully as they can be damaged by static charges. It is safer to leave them in their original packaging until you are ready to fit them. Even so, I can't recall having any problems with electrostatic charges.

Some of the simpler examples of layouts given show only the component side, usually because there are few or no breaks necessary in the trackside. The components could be mounted on the trackside; this method was chosen for the handbell circuit so that the insulated top side of the pcb backs on to the metal loudspeaker.

Notice that the layouts generally follow the pattern of the circuit diagrams. This makes it easier to trace the circuit through. Circuit conventions also help you to recognise standard blocks; inputs are usually on the left, outputs on the right. Note that the 0V rail generally runs along the bottom track and the + rail along the top.

4 Wiring

Single-strand copper wire with plastic insulation can be used between two fixed points of a circuit. However, where there is to be movement, a flexible connector or pointer for example, multi-strand insulated wire should be used. For link wires, the rule is the shorter the better to avoid stray pick-up.

The ends of the fine enamelled wire for the coils may need to be scraped clean before soldering. Although reels of this wire can be bought, if you strip an old transformer, the bobbin will generally supply more than is needed. The turns do not need to be layer wound. Scramble-winding by hand at one turn per second gives, by my calculator, 600 turns in ten minutes.

Take advantage of the modular circuit approach. Wire each stage separately and check it, if possible, before starting the next. One fault in a small circuit is far easier to find than two faults in a larger circuit.

Note that most projects have a large electrolytic capacitor from the + rail to the 0V. This is known as a decoupling capacitor. Its purpose is to short-circuit any signal voltage that may develop across the internal resistance of the battery. Such a signal could cause instability because the battery internal resistance is virtually in series with all stages and is therefore common to them. Similarly, care must be taken to keep all returns to the 0V line short to prevent any 'common-earth' loops.

5 Mechanical Construction

Most project chapters give mechanical details of boxes, keys, panels, etc., although sizes will often depend on the particular components to hand. However, a lot will depend on the intended application. Some of the music teaching aids, such as the Electronic Sol-Fa and the Rhythm Setter, could be useful as class demonstration models so could be larger.

Plastic project boxes in various sizes and many other components are advertised in the Maplin *Buyer's Guide to Electronic Components*, published yearly and available from the High Street newsagents. You may be lucky and have a local component shop that stocks many of the components that you need, or there are many other mail order electronic component suppliers and retail component shops up and down the country.

Chapter 2

BACK TO BASICS

If you are used to constructing practical electronic circuits and have some knowledge of musical theory you can skip this chapter and move on to sound-out the projects. For beginners, or those who need a refresher, this back-to-basics chapter offers some elementary theory on electronics and music.

1 Electronically Speaking

Electronics is all to do with *electrons* on the move.

Electrons are the negative parts of an atom, the smallest part of matter. You can't see them, but get them moving in a controlled way and you can do all kinds of interesting and useful things with them as we shall see.

Conductors are materials in which electrons can move easily; metals such as copper and aluminium are examples of very good conductors of electricity. You will need some copper wire to connect up these projects.

Current is the flow of electrons through conductors and is measured in amperes or amps.

Insulators are materials in which electronics do not move easily; for example, paper, wood and plastics. We use plastic covered wire to prevent currents flowing where wires can touch each other and cause short-circuits. Somewhere in between good conductors and insulators lie materials that offer resistance to a flow of current, like carbon (used for resistors); also the semiconductor materials for diodes and transistors, which we can use to control electron flow.

Resistance to a flow of current is measured in ohms. We use components that have resistance to current to perform useful functions. In this way, simple circuits can be constructed, for example, to produce and amplify musical sounds, to provide timing signals or to operate lights in a required sequence, etc.

Although a piece of wire is a conductor, an electric current will only flow through it if its ends are connected to something that gives electrical pressure.

Batteries are sources of electrical pressure. They consist of a cell or a number of cells.

Voltage is the term used for the electrical pressure of the battery. When a battery is charged, there are lots of electrons on the negative (−) terminal and lots of positive protons on the positive (+) terminal. The higher the voltage of the battery, the higher the current it will give when connected to a circuit. A **circuit** is the path through which the current flows. A battery that is not connected to anything has voltage, but no current passes through the air between its terminals because air is an insulator.

Don't do this, but if you connect a wire between the two terminals of a battery, electric current will flow through the wire — and quickly run your battery down! That's because the wire has only a small resistance to electric current.

It's not much use making an electric current flow through a wire if it doesn't do anything sensible. In fact, the wire does get hot, and if thin wire of a certain kind with more resistance is used, a smaller current will flow and the wire could serve as a heater. If a bulb is connected in the circuit, the thin wire (filament) inside a vacuum also gives light. The thin wire doesn't let so much current through and so the battery lasts longer. Special conductors such as carbon and tungsten have more resistance to electric currents.

Summing up, resistance is measured in ohms; pressure or voltage is measured in volts; current (usually indicated by I) is measured in amperes or amps for short.

You don't need to know this to make up the musical projects, but a useful law that relates these three quantities is known as Ohm's Law.

Voltage (volts) = Current (amps) × Resistance (ohms)

If two of these quantities are known, then the third can be calculated by applying Ohm's Law.

You can remember Ohm's Law by this triangle:

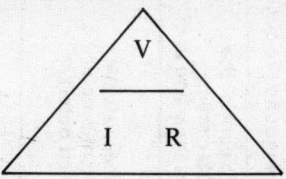

Cover up the quantity you need to know with your finger and it equals the other two: $V = I \times R$, $I = V/R$, $R = V/I$.

For instance, if a 6 volt battery is connected to a circuit of 12 ohms then the current flowing would be 0.5 amps, i.e.

Since $V = I \times R$, $I = V/R$,

so $I = 6/12 = \underline{0.5 \text{ amps}}$.

2 Components

Most of the following circuits will include electrical components such as resistors, capacitors, inductors or coils, transformers, loudspeakers, transistors, diodes, light-emitting diodes (LEDs), integrated circuits, switches, batteries.

Don't worry if some of this sounds like a foreign language to you.

The Electronic Circuit Component Chart (Figure 2.1) can help to identify components and to remind you of what they do. Refer to it as you read about the components.

Here are a few of the more common ones. You will learn more about how they are used as we go along.

Resistors, as mentioned, are used to allow a specific flow of current and generally have fixed values.

The electrons that flow around a circuit wire (and run the battery down very quickly) are reduced if a resistor is connected in circuit. In other words, the resistance reduces the current flowing according to its resistance, measured in ohms. High resistance values are measured in kilohms (1k = 1,000 ohms) and even Megohms (1M = 1,000,000 ohms). Fixed

Component	Appearance	Circuit symbol	Details
1. Battery		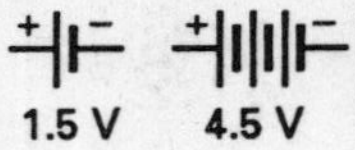 $+\|-$ 1.5 V $+\|\|\|\|-$ 4.5 V	A battery provides electrical energy. It gives out electrons at one terminal and takes them in at the other. The higher the voltage, the more electrons it can force around the circuit.
2. Resistor		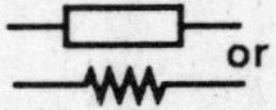 ▭ or ⌇⌇	Resistors are used to control the current flow around a circuit. Coloured bands give the value in ohms.
3. Variable Resistor		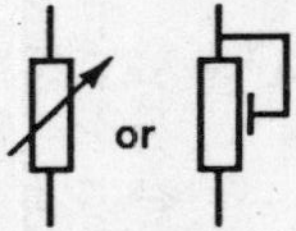 ▭ or ▭ with arrow	To vary the resistances, use two connections of a potentiometer (the middle tag and an outer)
4. Potential Divider		▭ with arrow down	A potential divider users all three connections of a potentiometer.

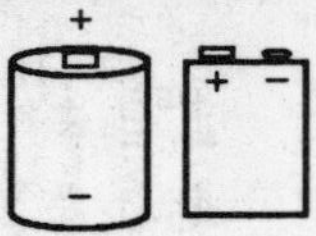

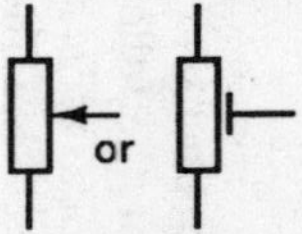

Fig. 2.1 Electronic Circuit Component Chart

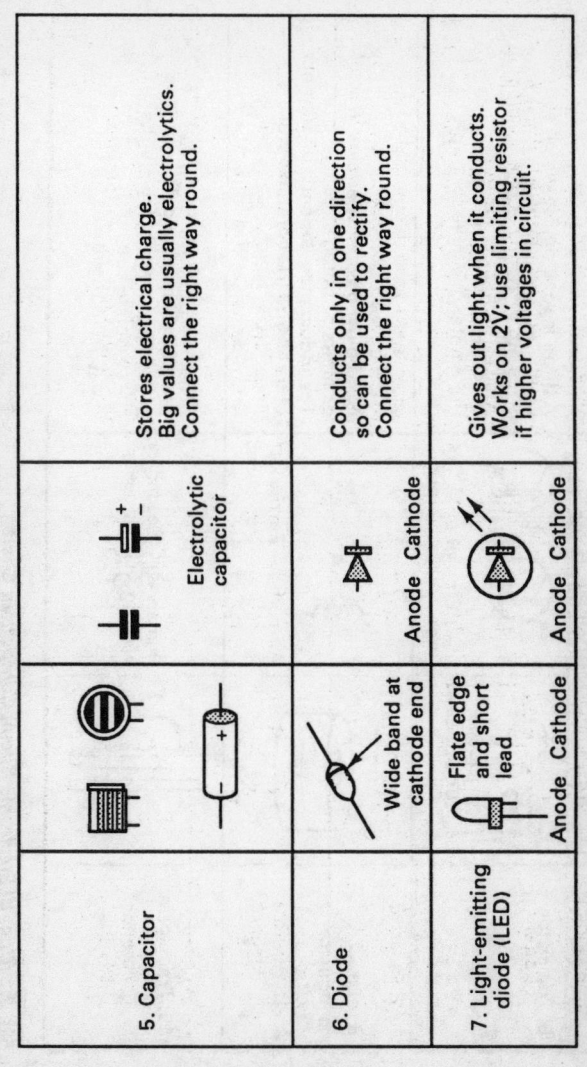

Fig. 2.1 (Cont'd) Electronic Circuit Component Chart

8. Transistor	 ![BC109 (under view)] b c e BC109 (under view)	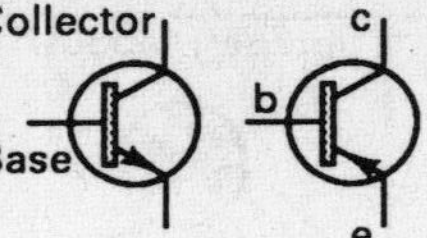 Collector c, Base b, Emitter e — npn; c, b, e — pnp	A three-legged device which can be used as an amplifier, oscillator or a switch. npn transistors work with the collector positive with respect to the emitter. (reverse for pnp).
9. Coil or inductor	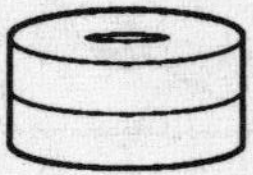	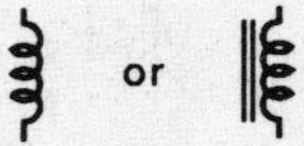 Air-cored or Ferrite- or iron-cored	Can be used with capacitors as filters or tuned circuits. The inductance depends on number of turns and the core material.
10. Transformer	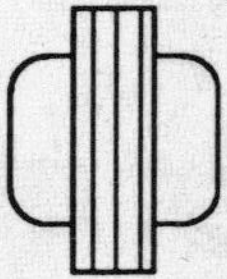	 Primary winding / Secondary winding	Transformers are a.c devices: – to step up/step down voltages, – to isolate circuits; e.g. from the mains.

Fig. 2.1 (Cont'd) Electronic Circuit Component Chart

resistors are usually colour-coded to indicate their values. It's as well to know the method of coding so that you can identify the different values for selection and when they are in circuit. You'll soon get to know the values when you've looked them up a few times!

The coding is generally in the form of colour bands around the body of the resistor. A group of bands bunched at one end indicate its value while a single band, gold or silver, indicates its tolerance (Gold ±5%; Silver ±10%).

For our purposes, all we need to know is that the colour of the first band represents the first figure of the value, the second band the second figure, and the third band indicates the number of noughts to follow.

The colour-code values are:

Black	0	0 × 1
Brown	1	1 × 10
Red	2	2 × 100
Orange	3	3 × 1000
Yellow	4	4 × 10,000
Green	5	5 × 100,000
Blue	6	6 × 1,000,000
Violet	7	7
Grey	8	8
White	9	9

Resistors are represented in a circuit by a rectangle or sometimes a 'zig-zag' as shown.

Variable resistors or potentiometers are used as controls, for instance as a volume control, and are represented by a rectangle with a short line close to it, or an arrow through it.

A capacitor stores electricity in a circuit. Early capacitors consisted of two metal plates separated by an air gap. Modern capacitors have metal foil plates rolled up with an insulating material sandwiched in between. If a battery is connected to the capacitor, a current will flow and gradually stop as an electrostatic charge builds up across the plates. This charge can be stored and then discharged into a suitable circuit to perform a useful function. Direct current (d.c.) cannot flow through a capacitor, so a capacitor can be used to block d.c. However, in a circuit with alternating currents (a.c.), for instance, sound wave signals, a capacitor can control the flow depending on its value, measured in microfarads (μF) or nanofarads (nF), and the frequency of the waveform flowing. Electrolytic capacitors are usually used for the bigger values of capacitance (identified by a + sign at one end) and must be connected the correct way round in a circuit. The values and maximum working voltages are usually written on capacitors.

Coils or inductors create a magnetic field in a circuit and store energy or control the flow of electricity depending on the value of the coil in henrys (H) or millihenrys (mH). Coils are used in some of the music projects together with a capacitor to make a tuned circuit for the tone generators. Each note can be tuned to a different pitch (frequency) by choosing a different value of capacitor across a coil. Alternatively, the frequency can be changed by varying the inductance of the coil (varying the number of turns or by putting a metallic or ferrite core in the coil).

Diodes are electronic one-way streets. They have high resistance in one direction and low resistance in the other. Therefore, current flows through a diode or rectifier in one direction only, so unlike a resistor, it must be connected the right way around. It is a twin leadout component with an anode and a cathode; a wide band at one end indicates the (+) cathode leadout. A more positive voltage at the anode end allows a current to flow through to the cathode. Diodes can be used to rectify alternating current to direct current or to isolate or to group circuits as in the Chordmaker project.

Transistors are at the heart of most circuits. These semiconductor devices can be used to switch, amplify and oscillate. The three connections are the emitter, the base and the collector. When measured for resistance, the transistor appears to be two diodes back-to-back. The BC109 used for some of the projects is an *npn*-type silicon transistor.

When used as an amplifier stage, small signal waveforms applied between emitter and base produce larger signals, hopefully of the same shape in the collector circuit. Similarly, when a transistor is used as an oscillator, a small current flowing in the base circuit produces a much larger current in the collector circuit. Now, if this output is fed back into the base in the correct phase it produces an even larger output until the circuit self-oscillates. The frequency of oscillation is determined by tuned circuits. In the music projects we use both resistance-capacitance (RC) and inductance-capacitance (LC) tuned circuits.

The layout diagrams for transistors show an underneath view of the leadout wires.

Integrated Circuits are a group of transistors manufactured on a small area of silicon crystal interconnected into a circuit. The use of ICs means fewer separate components and connections in a circuit. The dual in-line types used throughout these music projects include the CMOS 4011 quad 2-input NAND gate, used as a Schmitt trigger and an oscillator, the 1458 dual op-amp, used as a comparator, the NE555CP timer, the 4017 decade counter and the LM386N audio amplifier.

The layout diagrams for ICs show a top view of the pin connections.

3 Musically Speaking

All sounds are caused by vibrations of the air. If these vibrations are regular and continuous a musical sound is produced. You might argue about this definition when you consider some of the continuous sounds that are regularly inflicted on your ears.

However, whether we bow, blow, scrape, pluck or bang, we produce vibrations — the basic requirement for music.

Some form of musical notation is needed to tell us what sounds the composer has in mind; what notes to play, how loud to play them and for how long.

4 The Sound in Music
Musical sounds vary in many ways:

>height or depth, according to the number of air-vibrations in a given time, known as *pitch*;
>the relative length of the sound, or *note value*;
>the *strength* or loudness of the sound;
>accent or rhythm, the *time* in music, which is usually divided into bars to indicate the recurrent accent;
>speed or *tempo* at which music is played;
>the quality or *timbre* of the sound, harsh or pleasing, depending on the overtones and waveform produced by a voice or instrument.

5 Octaves and Scales
When the number of vibrations of a note in a given time double, the ear hears the same note but at a higher pitch. This note is said to be an *octave* higher. As the word suggests, there are eight notes within the compass of an octave. If you play the eight white notes on a piano starting on C (see Figure 7.1), you have the scale of C major. These notes belonging to the key are known as the *diatonic* notes. The sharps (#) and flats (b) in between are known as *chromatic* (coloured) notes. The key of C major has no sharps or flats, but as you will see from the Electronic Sol-Fa project, if you start the first note of a scale (keynote or Doh) on any other note except C, you will encounter sharps or flats in the major scale. For example, in the key of F (one flat) the fourth note is Bb, in the key of G (one sharp) the seventh note is F#. As there are 12 semitones in an octave and a scale can be formed on any of these, there are a lot of scales — too many to discuss here. Music is pitched in different keys, to add more interest.

6 The Piano
The piano covers a wide range of notes (7 octaves) so music for it is usually written on two lots of five lines placed above and below middle C. These are known as the Great Staff (see Figure 2.2). The pitch of the upper five lines of the staff is indicated by a treble or G clef (used also on the front panel

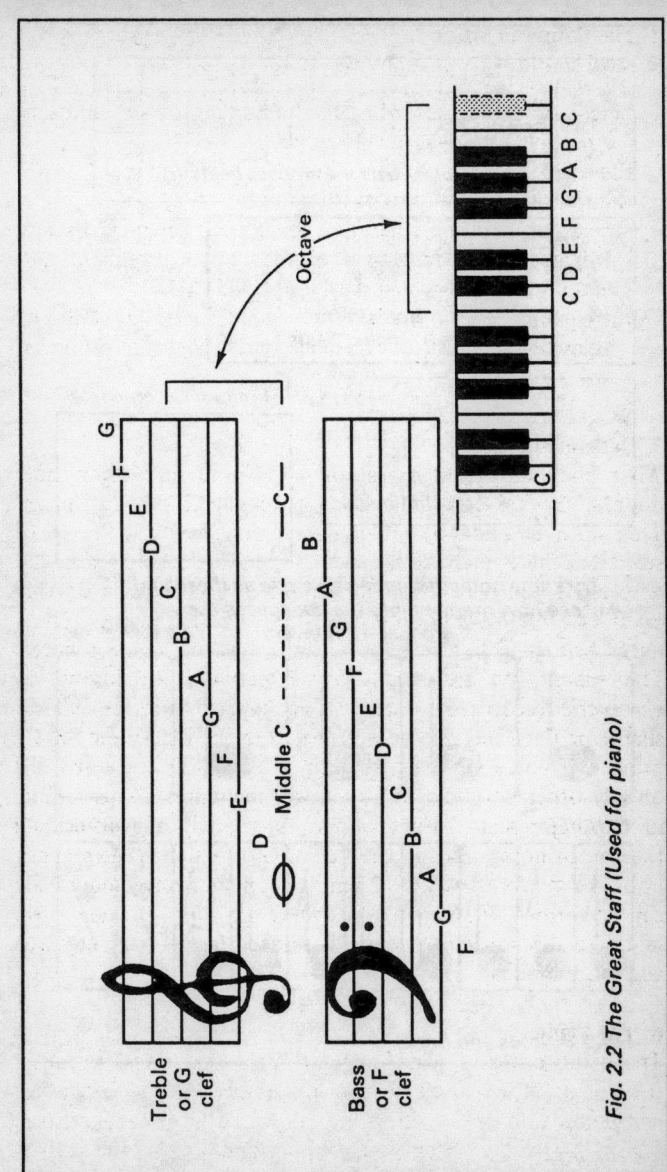

Fig. 2.2 The Great Staff (Used for piano)

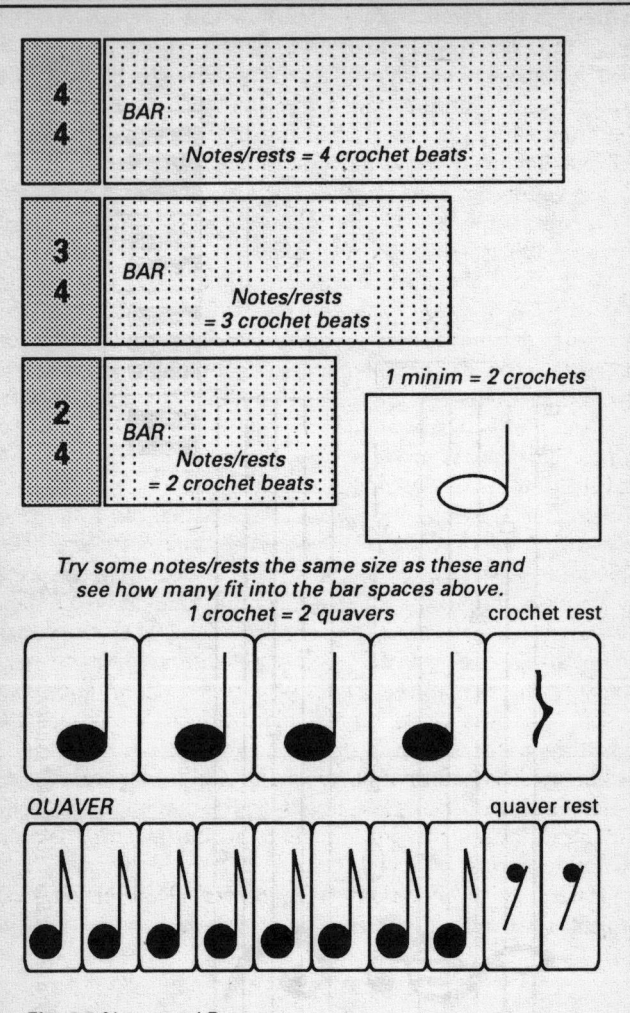

Fig. 2.3 Notes and Bars

diagram of the Chordmaker). To show this is the G clef, the twirl of the sign is wrapped around the second line G.

Instruments in medium pitch also use the treble clef. The lower five lines of the staff are known as the bass or F clef, and are used by the low-pitched instruments. Two octaves of a piano keyboard show how the notes relate to the music.

The various characteristics of musical sounds are exploited and further explained in the following projects. As rhythm and note values are discussed in the next chapter, Figure 2.3, Notes & Bars introduces a little musical theory. If you look at any page of music you will see that the width of a bar generally depends on how many notes it contains. For instance, writing a bar of four straight crochets needs less space than one with up to eight quavers (or sixteen semi-quavers) in it. Whatever the physical space taken up by the notes, the time taken to sing or play each note depends on its value and the speed or *tempo* of the music.

However, it helps to see musical notation in graph form, so in Figure 2.3 the widths of the bars, notes and rests have been shown in proportion to their relative time values. In this way, by simple addition, we see that a complete bar has to contain the correct value of notes and/or rests. This chart was originally A4-size and was used with cut-out notes to help young music students to grasp simple rhythms and time values. The idea could be extended to include semi-quavers — half the width of quavers.

For detailed information on major and minor scales, etc., there are a number of good books on the rudiments of music, and most instrument tutors include an introductory section on musical terms.

Time now to sound-out some of the electronic musical projects!

Chapter 3

RHYTHM CHECKER

The Rhythm Checker project provides a useful aid for music students to:

learn, or demonstrate, note and rest values;

appreciate time signatures and bar lengths;

play some basic musical rhythms.

For those new to constructing electronic projects, this chapter provides a gentle approach to circuit building, enabling the project to be tested at every stage so preventing multiple faults occurring — these can be tricky to find!

Fortunately, the rhythm checker project lends itself to this kind of approach. The indicator circuit can graduate from a simple series circuit of a battery and lamp (or LED) display, to a Schmitt trigger touch switch and astable multivibrator with sound and light output. The touch switch means that notes can be sounded by drawing a finger across the notes instead of a stylus; the body resistance supplies the contact.

For starters, beginners might try the Mark I indicator circuit after making up the bar frame.

1 Rhythm Checker Bar Frame (Figure 3.1)

This is a slide-rule type device in which plates representing musical notes (metallic) or rests (plastic) can be slotted to the value of a bar's length in the most common time signatures. The plate widths are proportional to their musical note (or rest) values, so together with the time signature, only the required number of plates fill the bar frame. Students can therefore use it to try out a problematic rhythm pattern, or to set up a new rhythm.

A printed version of the bar frame itself, used to teach simple rhythms, note values and bar lengths to students, is given in the previous chapter (Figure 2.3, Notes and Bars). An enlarged version of this can be made in card together with cut-out notes, etc.

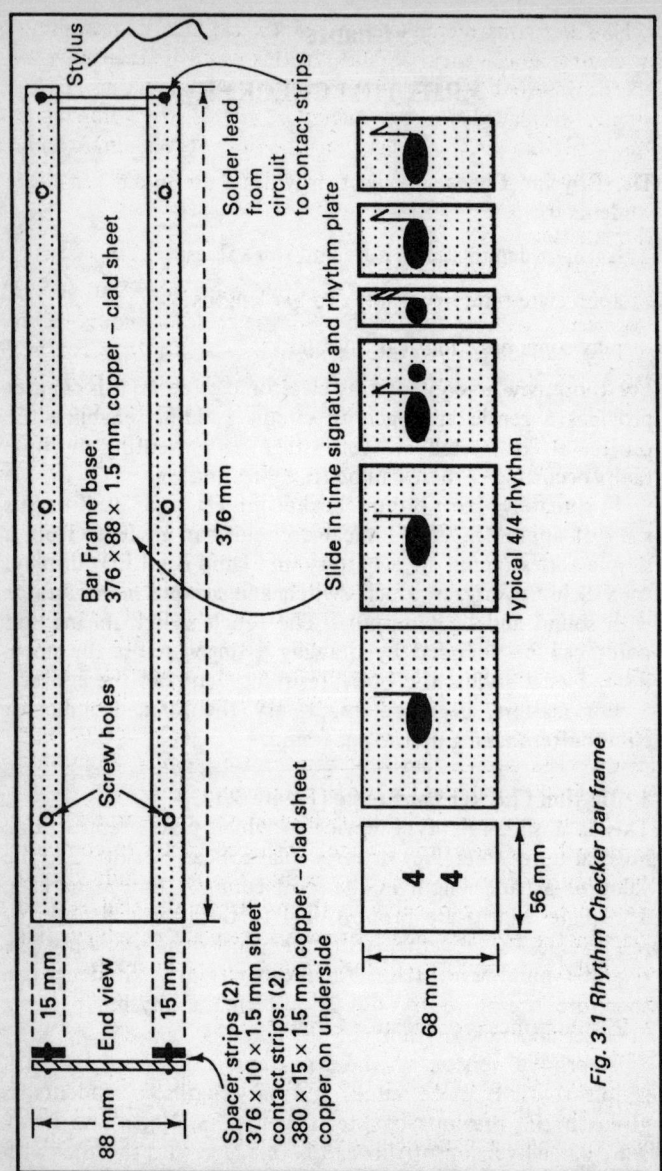

Fig. 3.1 Rhythm Checker bar frame

However, this metallic note version extends its usefulness by adding an electrical circuit. In this way, the relative note lengths inserted in the frame can be sounded out and/or visually displayed on an *indicator circuit*, by drawing a conductive stylus (or finger in the Mk II version) evenly across the bar frame.

Construction

The details for making the bar frame are shown in Figure 3.1. The bar frame base is shown as copper-clad sheet, but this is not essential. If copper-clad is used, the cladding should be on the top surface. However, the cladding on the top contact strips should be on the underside to make contact with the note-plates. Figure 3.2 shows the insert plates for the time signatures, notes and rests with dimensions. You will need one of each of the time signature plates, and a number of the notes and rests depending on the rhythms you want to try. The notes are cut from conductive metallic sheet, copper-clad board or aluminium, the width of each note being proportional to its musical value. For instance, a quaver is half the width of a crochet and a crochet is half the width of a minim. Likewise the equivalent rests are of the same widths as these notes, but are made from acrylic sheet. If copper-clad board is used for the notes, fewer blocks are needed as these can be reversed to indicate the equivalent rests on the acrylic side. However, as copper-clad tends to tarnish quickly, aluminium is a preferable material for the bar frame notes.

The notes must fit fairly tightly in the bar frame as they need to be in electrical contact with the frame to complete the circuit. For the stylus to sound the notes individually, there must be a gap between them. If copper-clad is used, chamfer the two side edges of each note-plate to achieve this, otherwise a narrow strip of insulating tape can be used.

2 Rhythm Checker Indicator Circuits
Mark I (Figure 3.3)
This simple circuit is for the beginner in electronics. Used with the rhythm frame, it provides a rhythmic light pattern when a hand-held stylus is drawn across the bar to show the relative musical time values. The three components

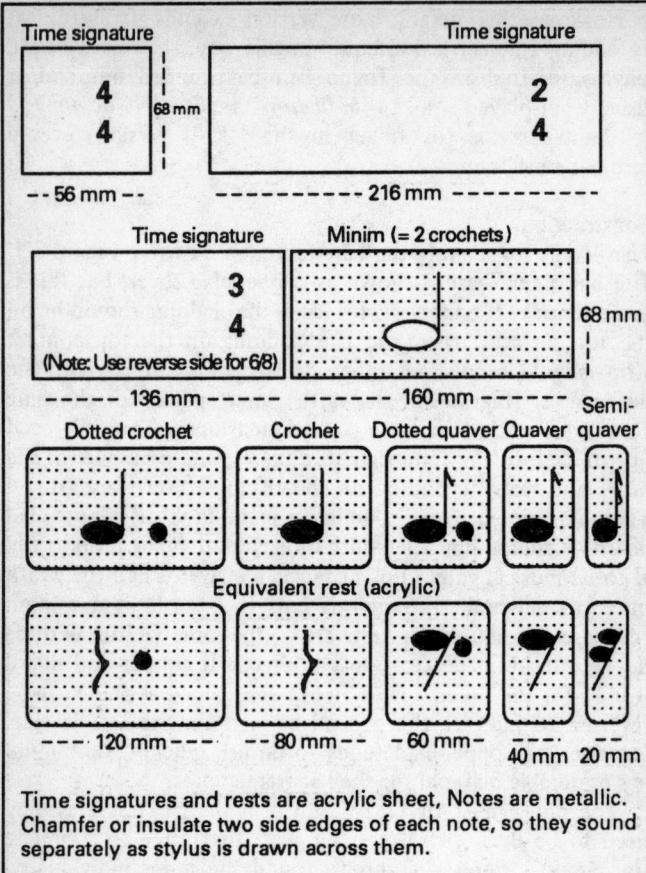

Fig. 3.2 Time signature, Notes and rests

can simply be wired together as indicated to prove the circuit. Alternatively, it is a good basic circuit to try laying out on stripboard. One such layout is suggested. Remember that an LED must be connected the right way round with its cathode (the leg on the flat side) towards the negative side of the battery. The limiting resistor connected in series with the

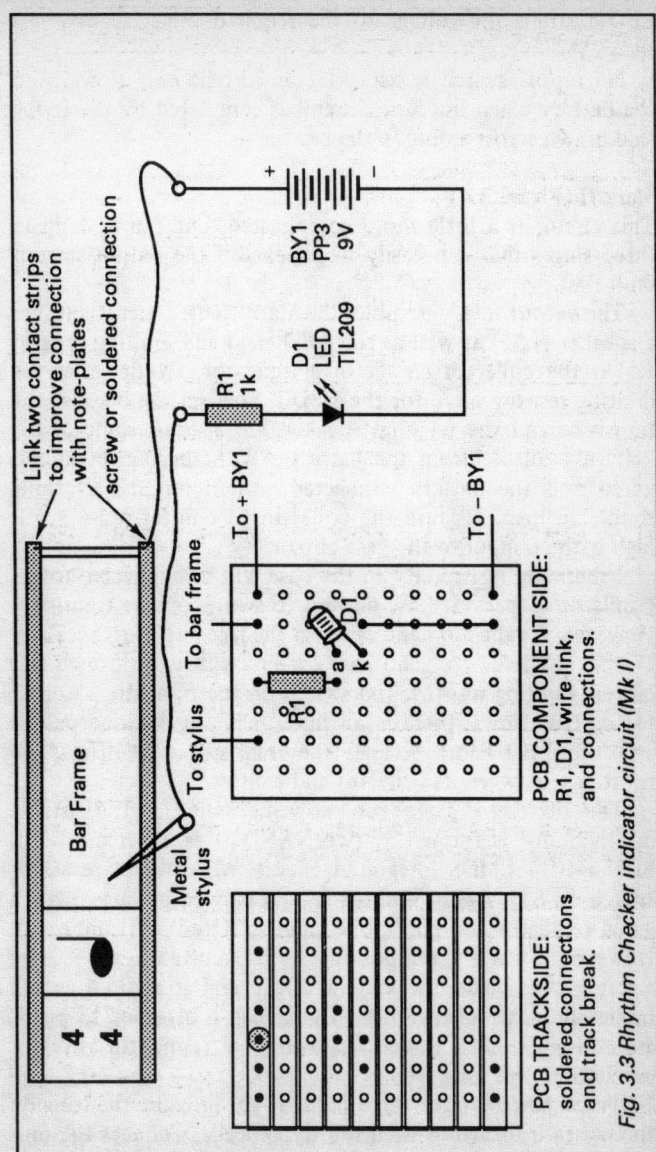

Fig. 3.3 Rhythm Checker indicator circuit (Mk I)

LED restricts the voltage to the required value, slightly less than 2V.

No on/off switch is necessary as current only flows from the battery when the series circuit is completed by the flying lead drawn across a note in the bar frame.

Mark II (Figure 3.4)

This circuit is a little more complicated, but can be built in three stages that can easily be checked if the output stage is built first.

The *output stage* resembles the Mark I series circuit, but has a speaker (LS1) as well as the LED (D1) and limiting resistor R5 in the collector circuit of a transistor. Notice that the limiting resistor value for the LED is now smaller because the transistor and the 64 ohm speaker will also use some of the voltage available from the battery. With the output circuit wired and the battery connected, switch on S1. Nothing should happen, because the collector to emitter resistance is high with no input to the base circuit.

Eventually resistor R4 in the base will be connected to the oscillator to provide base current to switch on the transistor. However, for the moment, connect the free end of this resistor to the +9V rail. A small base current will flow through this current-limiting resistor, and switch on the transistor. Collector current flows; there is an initial click in the loudspeaker and the LED lights because the transistor now offers low resistance between its collector and emitter.

The *oscillator stage* can now be wired. The astable multivibrator is formed by two NAND gates, IC1c and IC1d, of the 4011 CMOS integrated circuit. When these are wired, switch on S1. Again, nothing should happen because this is a gated oscillator and pin 9, one input to IC1c is not connected. However, if pin 9 is temporarily short-circuited to pin 8 using a screwdriver blade, the LED will light and an audio tone will be heard in the speaker. If a flying lead is attached to pin 9, and the +9V line is connected to the bar frame, the rhythms can be seen and sounded.

The *Schmitt trigger* stage IC1a, IC1b, provides the icing on the cake: it dispenses with the flying lead. Connect the pins of this stage as shown in the circuit diagram. As you can see,

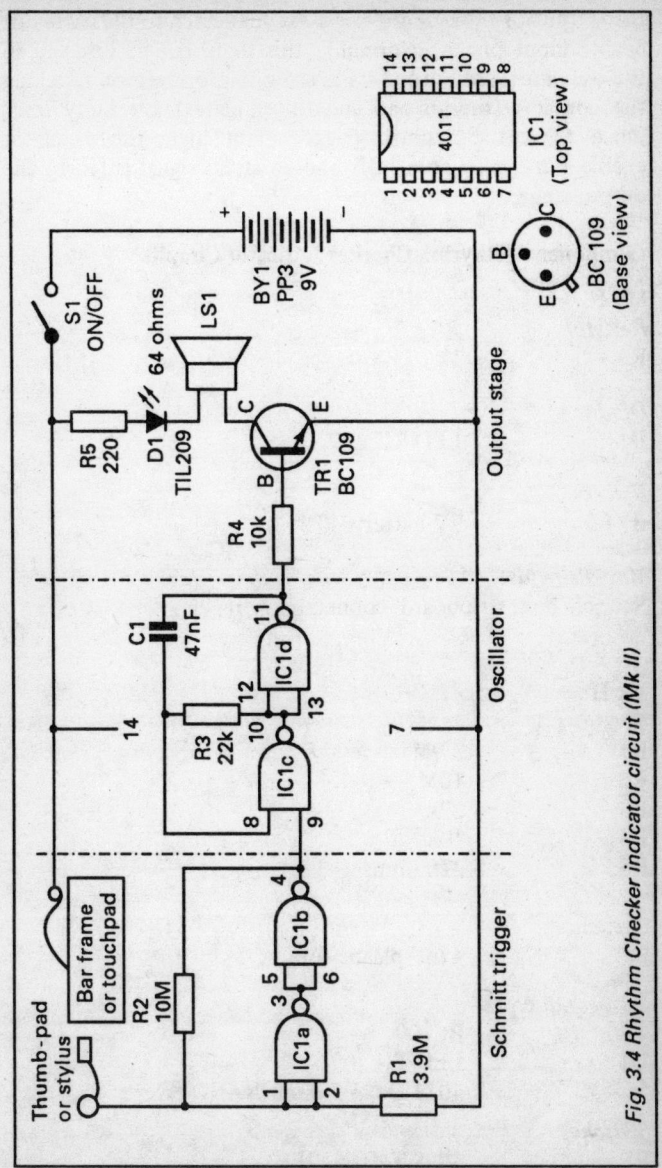

Fig. 3.4 Rhythm Checker indicator circuit (Mk II)

the output of the Schmitt trigger is connected to the oscillator enable input pin 9. Normally, this input is held low and so the oscillator and output stage are off. However, on touching the contacts (thumb pad and finger plates), the body resistance takes the Schmitt trigger input high, the oscillator enable pin 9 also goes high and an audio signal is fed to the output stage.

Components, Rhythm Checker Indicator Circuits

Mk I
Resistor
R1   1k

Diode
D1   LED TIL209

Battery
BY1   9V battery, PP3

Miscellaneous
Suitable box, stripboard, connecting wire, etc.

Mk II
Resistors
R1   3.9M
R2   10M
R3   22k
R4   10k
R5   220 ohms

Capacitors
C1   47nF plastic foil

Semiconductors
TR1   BC109
D1   LED TIL209
IC1   4011 Quad 2-input NAND-gate

Switches
S1   S.P.S.T. (ON/OFF)

Loudspeaker
LS1   64 ohms

Battery
BY1   9V, PP3

Miscellaneous
Suitable case, stripboard, connecting wire, etc.

The rhythm checker indicator circuit offers more flexibility if it is housed as a separate unit because there are some useful variations to the bar frame, as follows:

3 Rhythm Checker Touchpad (Figure 3.5)

The indicator circuit may also be connected to the base of this simple touchpad by the two touch wires. The diagram Figure 3.5 shows a paper or plastic overlay template with cut-out holes for the various rhythms. An enlarged version of this overlay can be clipped (or glued) to a metallic or foil-covered baseplate (baking foil glued to hardboard works well, but do get permission from the cook!). The hole patterns are designed to familiarise students with note values and time signatures. Relative note values are given from a semibreve down to 16 semiquavers performed in the same length of time.

The metallised baseplate is connected to the circuit in the same way as the bar frame. Twin connecting wire is needed if the thumbpad is mounted on the touchpad. The thumbpad terminal must be isolated from the metal baseplate.

A sharp knife is needed to cut out the holes in the overlay for the notes. The figure can be reproduced, preferably enlarged on a photocopier. The original was A4 size to accommodate big fingers, but if overlap of notes occurs, drawing the end of a paper clip across the notes does the trick. If an A4-size overlay is used, the baseplate should be slightly larger, say 12 in × 9 in.

4 Finger-tapping Rhythmboard (Figure 3.6)

The finger-tapping rhythmboard can either be built separately on its own baseplate, or as just another overlay to clip on to the baseplate of the rhythm checker touchpad described above.

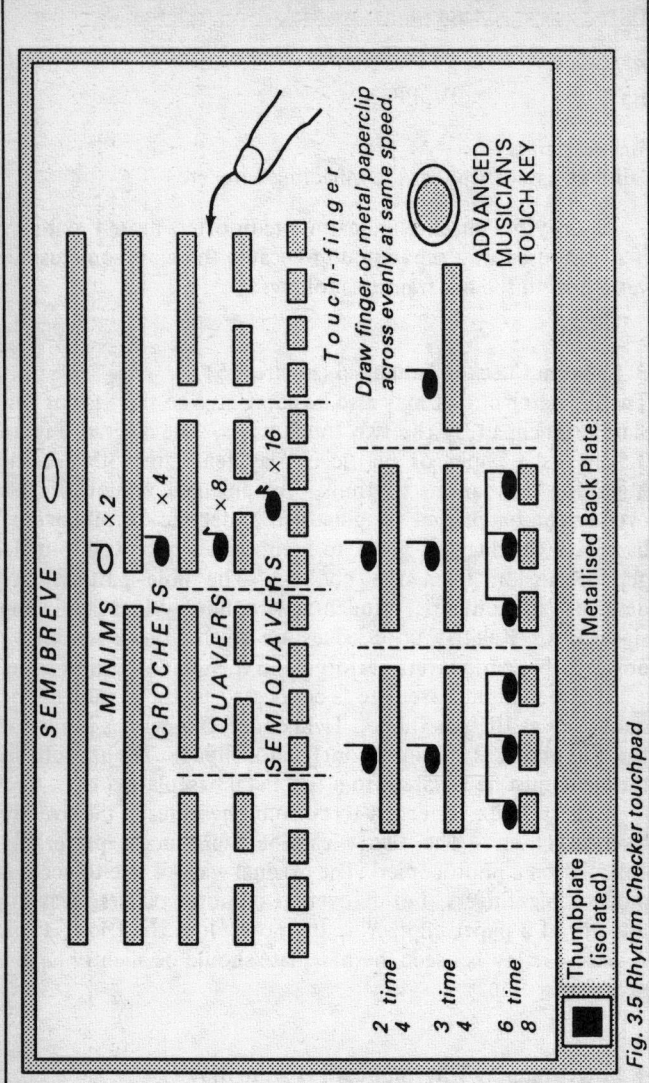

Fig. 3.5 Rhythm Checker touchpad

Fig. 3.6 Finger-tapping rhythmboard

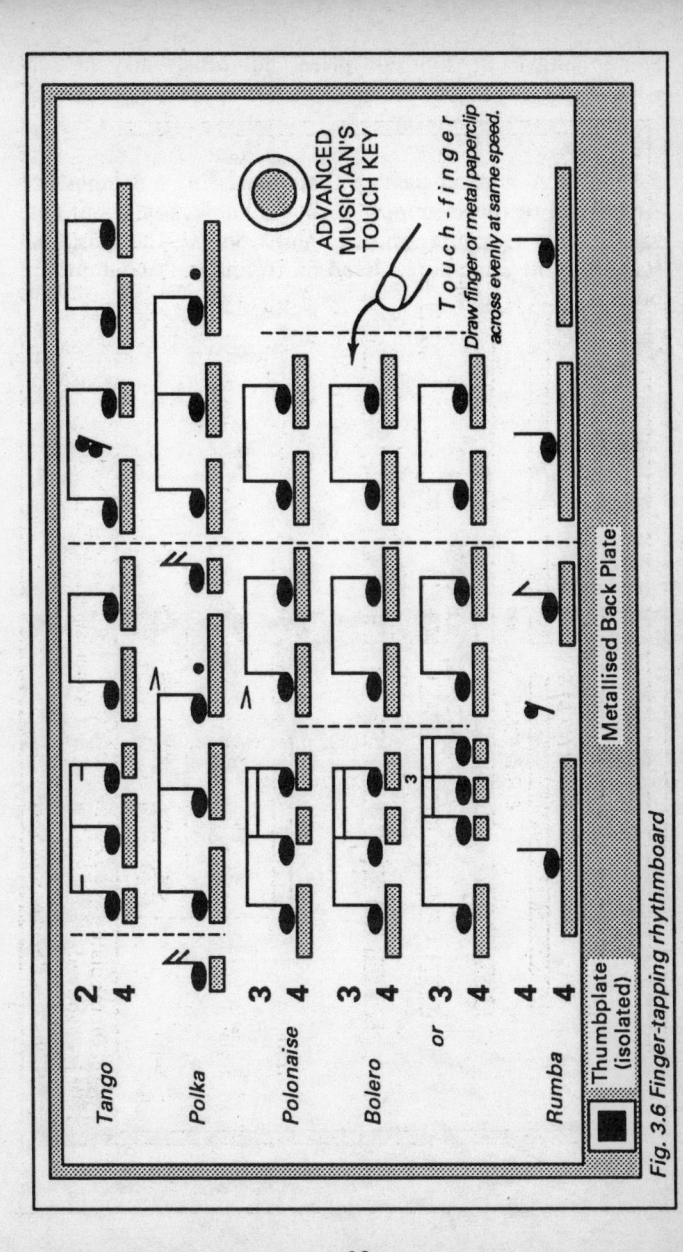

Some familiar rhythms are given, but others may be substituted or tapped out on the advanced musician's touchkey.

5 Coda: A musical passage that rounds off a composition
At the end of this chapter you should know more about note values, time signatures and rhythms. So far, the music has stayed on the same note. Read on to find out about musical pitch.

Chapter 4

ELECTRONIC SOL-FA

This project provides a useful aid for sight-singing, teaching and learning intervals, and the names of the notes in a musical scale.

The Tonic Sol-fa method of teaching singing was widely used in the past and there is still a lot of vocal music around with this doh, ray, me notation added.

The problem with the traditional staff notation (A, B, C, D, E, F, G) is that the notes are at a fixed pitch or frequency. It is fairly easy to find them on a musical instrument, but more difficult for sight-singers when the music is in a key that has several sharps or flats. And if a tune is written in a different key, it means learning an entirely new set of names for the notes.

However, with the Tonic Sol-fa notation the relative notes, whatever the key, are simply Doh, Ray, Me, Fah, Soh, Lah, Te, Doh. Remember Maria and the Von Trapp family in 'The Sound of Music'? And the last note of a melody, as Julie Andrews reminds us, brings us back to Doh − well usually!

In addition to the Doh, Ray, Me . . . notes on the slide, the scientific names are also given as a bonus: tonic, supertonic, mediant, etc. These will be found useful for music students in the study of harmony.

Tonic Sol-fa is referred to as a *movable* Doh system, so the Electronic Sol-fa system has been designed to be just that! The fixed staff notes are fixed on the stock of a slide-rule type scale and the movable tonic sol-fa notes can be moved up or down according to the chosen key signature. With Doh opposite the desired keynote, the fixed notes can be sounded individually to produce scales and intervals in the selected key.

If this sounds complicated, the principle is easily seen by reference to Figure 4.1.

1 Mechanical Construction (Figure 4.1)
The size of the electronic sol-fa unit is primarily dictated by

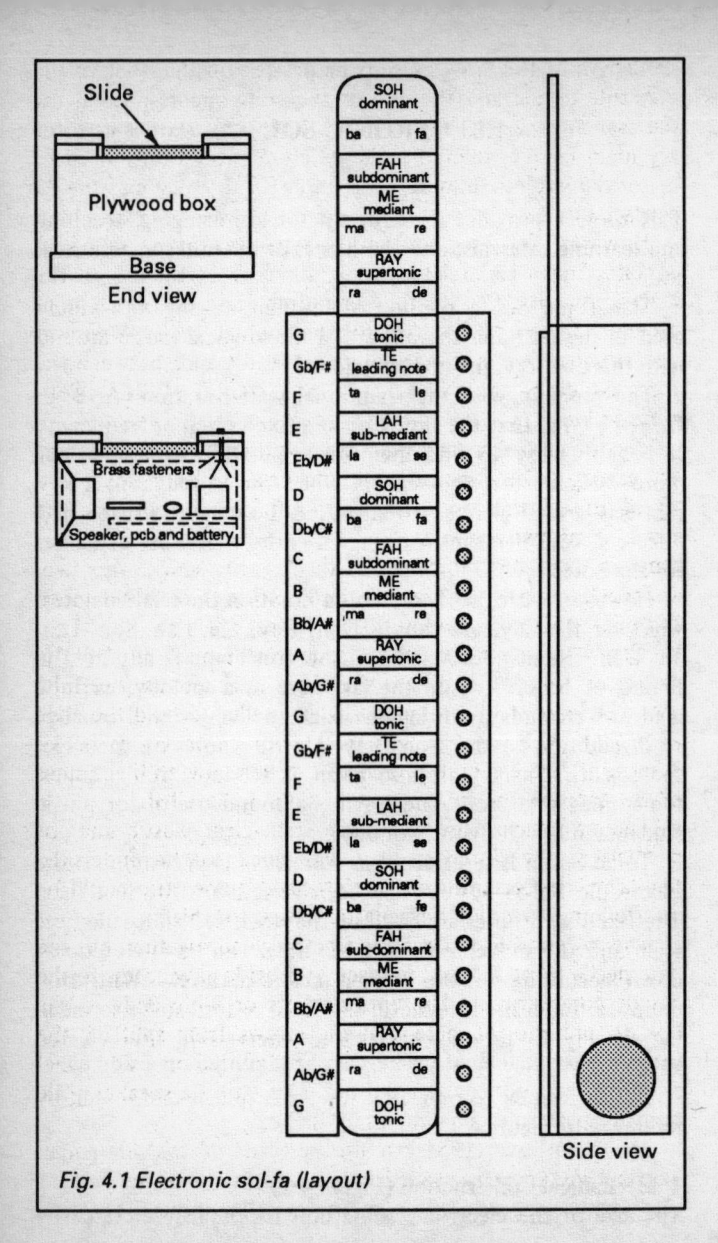

Fig. 4.1 Electronic sol-fa (layout)

the length of the two-octave scale printed on the stock of the slide-rule device and the size of the components needed for the oscillator circuit (Figure 4.2). If a separate box is used for the main circuit, then the slide-rule unit would only need to house the pitch-determining resistors (R8 to R31) and the 25 contact studs.

However, for ease of working, the prototype was self-contained in a roomy 3-ply box, about twice the size of the illustration. Brass paper fasteners placed in a row of 25 holes drilled 1cm apart were used for the contact studs. Whatever size is chosen, it is important that the distance between the studs should be equal and match the distances of the sections on the slide.

The original lettering was done with rub-down transfers, but can easily be done on computer or as a photocopy blow-up of the illustration. The staff notation lettering on the stock spans two octaves ($G_1 - G - G^1$), which covers the range of most vocal melodies. However, the slide covers two-and-a-half octaves (so that we don't run out of notes when moving the slide to other keys).

The two guide strips for the slide can be made from 3-ply, or better still, from plastic the same thickness as the slide. Two retaining strips are made slightly wider to hold the slide in place. Alternatively, providing your carpentry is better than mine, these retainers could be dispensed with if the sides of the slide and the guide strips are suitably bevelled.

The prototype slide was made from clear plastic and for protection, the lettering strip was glued to the underside. Similarly, the lettering on the left-hand guide strip could be inserted underneath if clear plastic is used for this.

The plywood box can be tacked or glued together, but the base needs to be screwed in place to provide easy access to the circuit components. Use corner fillets or 5-ply ends to the box to prevent the base-retaining screws from splitting the wood. The small loudspeaker can be mounted on a side panel as convenient; either use a cut-out covered by a speaker grille or drill a pattern of small holes in the box.

The on/off switch S1 can also be fitted on the side panel. A small hole must be drilled in the bottom end of the box for the flexible lead connected to the metallic stylus. The lead

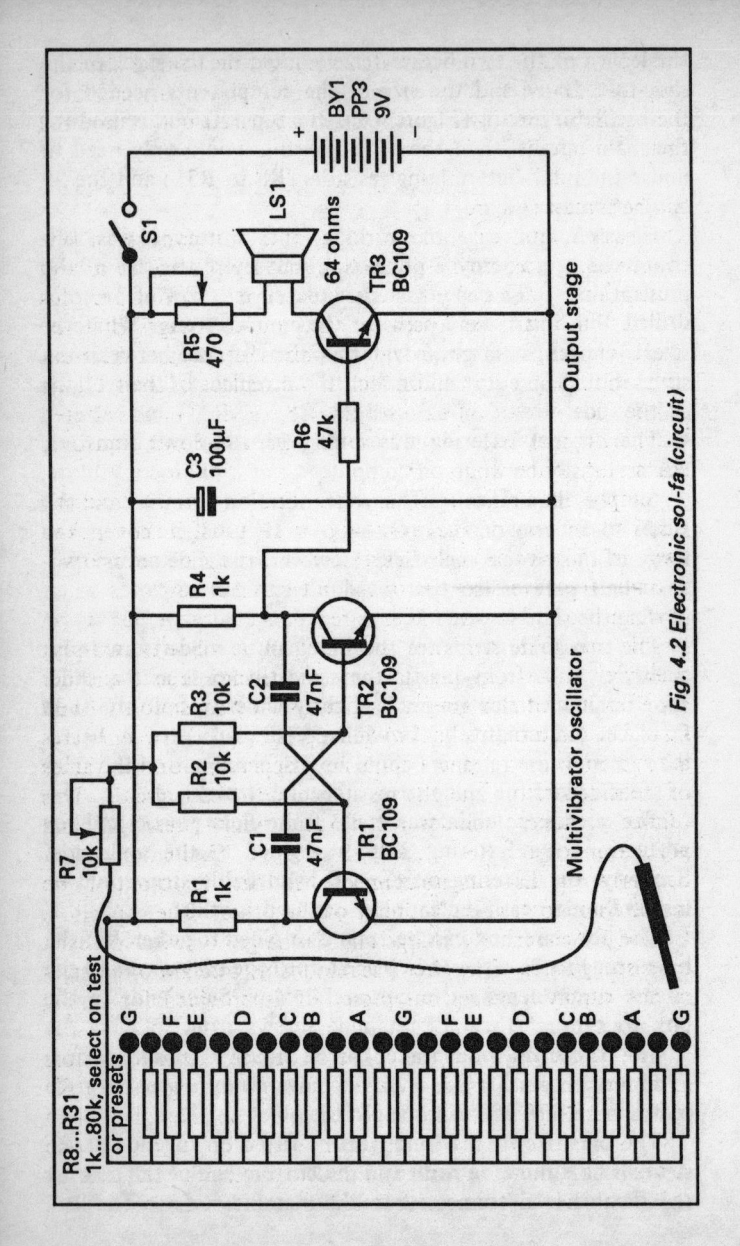

Fig. 4.2 Electronic sol-fa (circuit)

should be multi-strand insulated wire for mechanical strength and to allow flexibility of movement. The stylus can be a meter probe or a piece of brass rod finished with a pointed end.

2 The Circuit (Figure 4.2)

The basic circuit of the electronic sol-fa unit consists of two parts, namely, an oscillator feeding an output stage to drive a loudspeaker. The rhythm checker circuit in the previous chapter also had an oscillator and output stage, but the electronic sol-fa circuit provides a means of selecting different notes by using a stylus. In fact, if we replaced R3 of Figure 3.4 by our note-forming resistors (R8 . . . R31) and selected them with a stylus we could use that circuit for our electronic sol-fa unit. Then why not?

Simply to demonstrate another kind of circuit (and it's easier to understand the working of a circuit if it isn't tucked away in a black box) the oscillator uses discrete components — two *npn* transistors instead of the integrated circuit.

Transistors TR1 and TR2 are connected together as an astable multivibrator, i.e. a circuit that vibrates without stability. It stays in one state for a time then flips to another state because of changing voltages caused by capacitors C1 and C2. The circuit flips backwards and forwards between states ad infinitum at a speed (frequency) depending on the values of these capacitors and the resistance in the base circuit. This tuning resistance includes R2, R3, the tuning preset R7, and whichever resistor R8 . . . R31 is selected. Oscillations occur because the transistors are connected so that the output of one is fed back to trigger the input of the other. The capacitors C1 and C2 alternately charge and discharge to switch on each transistor in turn. The circuit is reasonably tolerant of changes in the supply voltage, but preset R7 is provided for tuning purposes.

The square pulses at the collector of one of the transistors (TR2 in this case) are fed via the current-limiting resistor R6 to the base of TR3. The amplified signal on the collector of TR3 is applied to the loudspeaker LS1. Preset resistor R5 serves as a volume control and also to extend battery life by reducing the collector current of output transistor TR3.

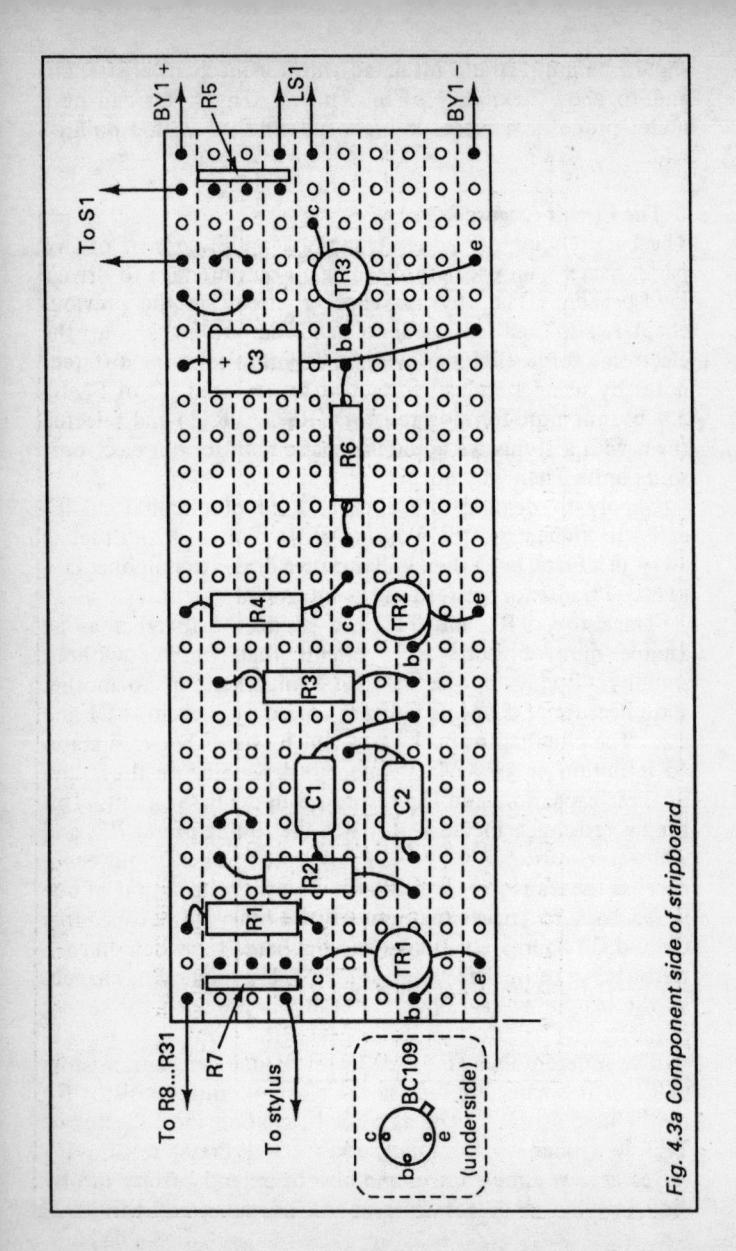

Fig. 4.3a Component side of stripboard

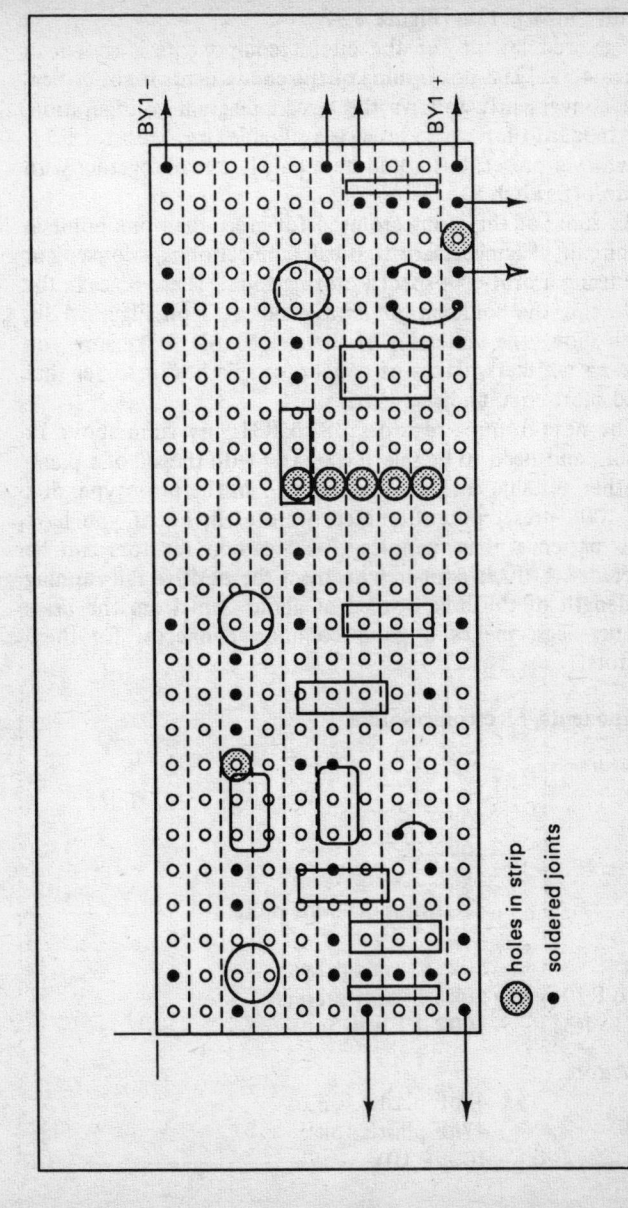

Fig. 4.3b Underside of stripboard

Circuit Construction (Figure 4.3)

A suggested layout for the circuit components is shown in Figure 4.3a. The positioning of the components is not critical so it conveniently follows the circuit diagram configuration. A stripboard of 10 × 25 holes is a suitable size. Resistor R5 is shown as a preset, but could be a panel control together with the on/off switch S1.

As some of the strips are used for more than one point in the circuit, it is necessary to break connection at a convenient hole using a drill or a special cutting tool. These holes in the strips and the soldered joints are indicated on Figure 4.3b, which shows the underside of the stripboard. Make sure you leave no whiskers, flakes of copper, or splashes of solder that could short-circuit adjacent strips.

The note-forming resistors R8 to R31 vary from about 1k to 80k, and need to be selected on test with the aid of a piano or other suitable musical instrument. In the prototype, 50k and 100k preset potentiometers were used, but if you have more patience than pennies, fixed carbon resistors can be selected. A thick copper wire from the positive rail running the length of the box, spaced at about 2cm from the brass fastener legs, makes a good common connector for these resistors.

Components, Electronic Sol-Fa

Resistors

R1	1k
R2	10k
R3	10k
R4	1k
R5	470 ohms 0.1 watt preset
R6	47k
R7	10k 0.1 watt preset
R8 to R20	50k 0.1 watt presets
R21 to R31	100k 0.1 watt presets

Capacitors

C1	47nF plastic foil
C2	47nF plastic foil
C3	100μF 10V

Semiconductors
TR1   BC109
TR2   BC109
TR3   BC109

Switches
S1   S.P.S.T.

Loudspeaker
LS1   Miniature type having an impedance in the range 25 to 80 ohms

Miscellaneous
Case, component board, 9 volt battery, connector to suit, wire, 25 brass paper fasteners, etc.

3 Tuning

Start tuning from the highest G — the top stud without a resistor. Touch the stylus on to this stud, and adjust R7 until the note from the unit corresponds to a G on your musical instrument. As you tune the note in towards your standard note, you will hear a slow beating sound which corresponds to the difference frequency between the two notes. Tune until this beat slows down to a stop when the two notes are in unison. With the 24 presets, it is then possible to tune down a semitone at a time, using your standard, throughout the two-octave range. When all presets (or resistors) are set, the unit is ready for use.

The unit is of course monophonic: i.e. it can sound only one note at a time. Intervals between two musical notes can only be sounded in sequence, but if there is a need to sound these notes simultaneously, then two identical circuits can be incorporated. These can share the same power supply and tuning resistors R8 to R31.

Chapter 5

DYNAMICS RANGER

Having dealt with two of the characteristics of musical sounds, note length and pitch, we now consider loudness. This is probably the easiest to understand, but the most difficult to define. If a tuning fork, musical instrument or your vocal cords vibrate at 440 times per second (440Hz in electronic terms), the pitch is naturally an A in anybody's book, but loudness is another story.

Force marks for intensity in music generally range from pp (pianissimo = very soft) through to f (forte = loud) and ff. But how loud is forte? To the sensitive soul on the front row of the concert hall — too loud! There should be a law against it, and there is one for sounds and horns of a different kind out-of-doors:

'the offending sound should be limited to 92 decibels at a distance of not less than 5.2 metres from the nearest point to the carriageway'.

The *Concise Oxford Dictionary of Music* dismisses the term forte as 'strong, i.e. loud', but how loud is loud? To musicians, loudness is subjective and depends on many variables. In electronics we need accurate assessments of sound intensity and sound level meters are used that simulate the response of the human ear. These are calibrated in decibels (dB), 0dB is the threshold of hearing and 1dB is the smallest *change* detectable by the ear. More familiar is the simple type of sound level meter we find in tape recorders and music centres.

The volume range of a symphony orchestra may vary from 30dB to 110dB depending on the number and type of instruments, the dynamics scored (pp, mf, ff, etc.) and the distance from the listener. If our ears can only stand about 130dB, then with a stentorian trombonist in full flight emitting 80dB you don't need to be a maths genius to realise that decibels just don't add up, otherwise we would be permanently deafened by a trio of trombones. In fact, two sounds of equal loudness give an overall sound that is only 3dB higher than one of them. And you need four instruments or voices sounding at equal loudness to double the volume (a 6dB increase). That's why doubling the number of voices in a choir doesn't double the volume.

As you will realise, the musical terms for loudness are relative, and a pianissimo or a forte, which can be produced by one or a number of different instruments and varies with the mood of the music, defies being labelled as so many decibels. For this reason, the Dynamics Ranger, as shown on the front panel diagram Figure 5.1, only indicates comparative loudness.

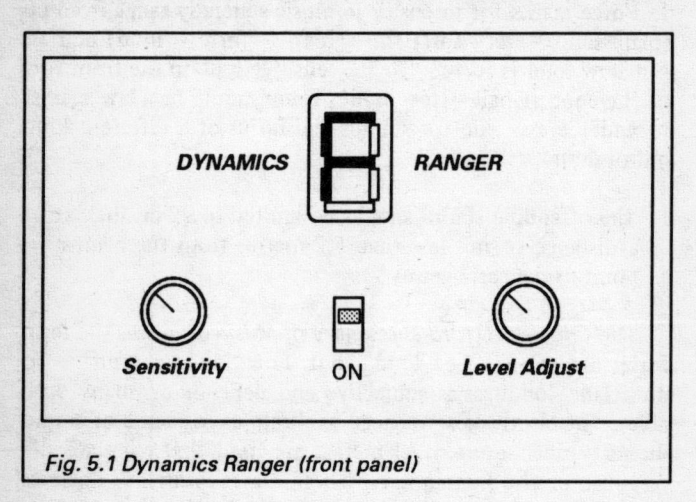

Fig. 5.1 Dynamics Ranger (front panel)

1 Circuit (Figure 5.2)

As musical loudness terms are comparative, what better circuit blocks to use than comparators! The basic circuit of the Dynamics Ranger consists of a microphone amplifier followed by a bank of four comparator circuits, the outputs wired so that four LED segments of a 7-segment LED glow progressively in sequence depending on the sound level. Individual LEDs could be used instead of the 7-segment LED, but the novelty of the present circuit arrangement is that on a loud

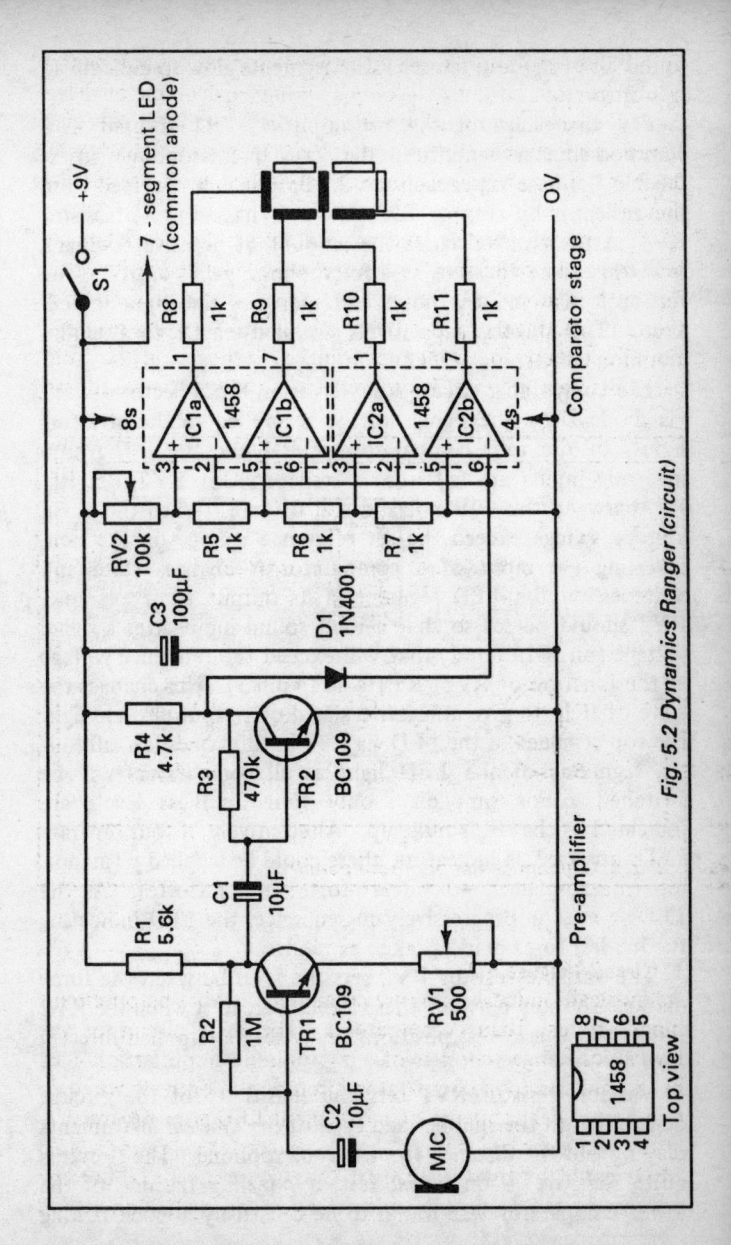

Fig. 5.2 Dynamics Ranger (circuit)

sound all of the four chosen LED segments glow to indicate an 'F' for forte.

The first stage of the pre-amplifier, TR1 is used as a common-emitter amplifier, the crystal microphone insert feeding the base via capacitor C2. Base bias is provided from the collector by resistor R2. The unbypassed preset resistor RV1 in the emitter varies the amount of negative feedback over this stage to serve as a microphone gain control. The feedback also reduces noise and increases the input impedance. This stage is capacitively coupled via C1 to a similar common-emitter stage TR2.

The comparator circuit is driven from the collector of TR2 via diode D1. The signal voltage is applied to the inverting inputs of the four comparators of IC1 and IC2. The non-inverting inputs are fed from a resistor chain, RV2, R5, R6, R7, between the +9V rail and the 0V rail. When the signal voltage swings exceed the set reference voltage on the non-inverting (+) input of a comparator it changes state and switches on the LED segment in its output. For example, RV2 should be set so that a loud sound input gives a signal voltage (on IC1a pin 2) that will exceed the reference voltage at the junction of RV2–R5 (to IC1a pin 3). This changes the state of IC1a to give a negative signal on output pin 1 to light the top segment of the LED via R8. In this condition, all four 'F' segments of the LED light as all comparators will be switched. For simplicity, only four loudness levels are indicated as the 'F' is built up. Alternatively, if four separate LEDs are used as indicators, these could be labelled p (piano), mp (mezzo-piano), mf (mezzo-forte) and f (forte). As the LEDs come on progressively in sequence, the LED indicating the loudest force mark is taken as valid.

The variable resistor RV2 sets the level between the force marks. The step between these levels is greatest when the RV2 slider is at the lowest position: i.e. with +9V applied directly to IC1a pin 3.

Variable resistor RV1 sets the sensitivity of the circuit, depending on the number and type of the musical instruments playing and the distance from the microphone. The circuit is quite sensitive — on initial test, a puzzling flashing of the lower comparator was found to be caused by a clock ticking six feet away. The circuit worked best with a crystal insert, but was still reasonably effective with a fairly low-impedance electro-magnetic insert. Note that the crystal microphone is a high impedance device and, to avoid unwanted hum and noise pick-up, should not be connected via a long cable.

The electrolytic capacitor C3 is included for decoupling purposes to prevent any tendency for the circuit to oscillate.

2 Construction

The circuit is easy to construct on a piece of stripboard (40 × 15 holes) using the methods described in earlier projects. There are no particular restrictions except that connecting leads, particularly in the amplifier front end, should be kept as short as possible to avoid noise pick-up which could give false readings.

The comparator section can easily be tested on its own by disconnecting D1 and applying a varying voltage between the input lead to the inverting inputs and 0V. The LEDs will glow progressively in sequence as the input voltage increases. A resistor value in the reference voltage chain (R5, R6, R7) can be altered slightly to vary the step between two levels if necessary. Increase the value to widen a step, decrease the value to reduce the step.

As indicated, two 1458 dual op-amps are needed to provide the four comparators. Make sure that the +9V supply is connected to pin 8 and the 0V to pin 4 of both chips. Alternatively, one 339 quad comparator could be used instead of two 1458 chips as the 339 packs four independent comparators.

The power requirements of the Dynamics Ranger are modest. The quiescent current consumption, no LED segments lit, was 6mA. When a loud signal was applied, this current increased to about 20mA.

A small plastic project box, 58 × 123.5 × 40.5mm (Maplin ABS Box PX2) with a plastic lid is suitable for housing this project. The front cover needs to be drilled to take the Sensitivity and Level Adjust potentiometers and the on/off switch. The on/off switch can be on one of these rotary potentiometers if desired. The 7-segment LED is also mounted on the front panel. A suitable 0.5-in display with common anode is obtainable from Maplin, order code FR39N.

Components, Dynamics Ranger

Resistors

R1	5.6k
R2	1M
R3	470k
R4	4.7k
R5	1k
R6	1k
R7	1k
R8	1k
R9	1k
R10	1k
R11	1k

Potentiometers

RV1	500 ohm potentiometer (Sensitivity)
RV2	100k potentiometer (Level Adjust)

Capacitors

C1	10µF 10V elect
C2	10µF 10V elect
C3	100µF 10V elect

Semiconductors

TR1	BC109
TR2	BC109
IC1	1458 Dual op-amp
IC2	1458 Dual op-amp
D1	1N4001

7-segment LED (common anode, see text)

Switches

S1	S.P.S.T.

Miscellaneous
Crystal microphone insert, case, stripboard, 9 volt battery, connector to suit, wire, etc.

Chapter 6

TUNE-UP BOX

A set of pitch-pipes is a useful accessory to anyone who trains or conducts choirs. Pianos are not always conveniently to hand, and to pitch the notes of the initial chord for unaccompanied singing can be something of a problem. Somehow, the notes you sound on the piano and the notes you sing to the choir on your return after negotiating a few music stands and a row or two of chattering singers are not always the same. And the tenors will be quick to point this out, if the resulting pitch change has not permanently damaged their vocal cords. Usually, tenors don't like very high notes, and basses don't like very low notes, so if you're not careful you could have a 'pitch battle' on your hands.

There are other good reasons for producing your own notes. I well remember an embarrassing moment in the world-renowned Amsterdam Concertgebouw, no less. After asking the Dutch pianist for an '*Ay*', the Dutch sound for an 'E', I launched the choir into a song, only to discover that they had started on 'A'. He had assumed I had asked for it in English — take 2! Not many musicians are blessed with perfect pitch so a compact, electronic equivalent of traditional pitch-pipes could find wide application, especially for amateur musicians.

1 Construction (Figure 6.1)

As can be seen from the suggested layout, the components are few and can be easily mounted on a small piece of stripboard, say 1in × 2in. The example shows only the underside of the stripboard with the solder points, and the breaks in the copper strip under the IC. However, for clarity, the components on the other side of the stripboard have been labelled and shown angled towards their respective solder points.

On the front layout, the three potentiometers are shown panel-mounted, but the two presets, 'Volume' and 'A Adjust', could be skeleton potentiometers mounted on the edge of the stripboard. Note the connections to the rear of RV1 and R3.

Fig. 6.1 Tune-up box (layout)

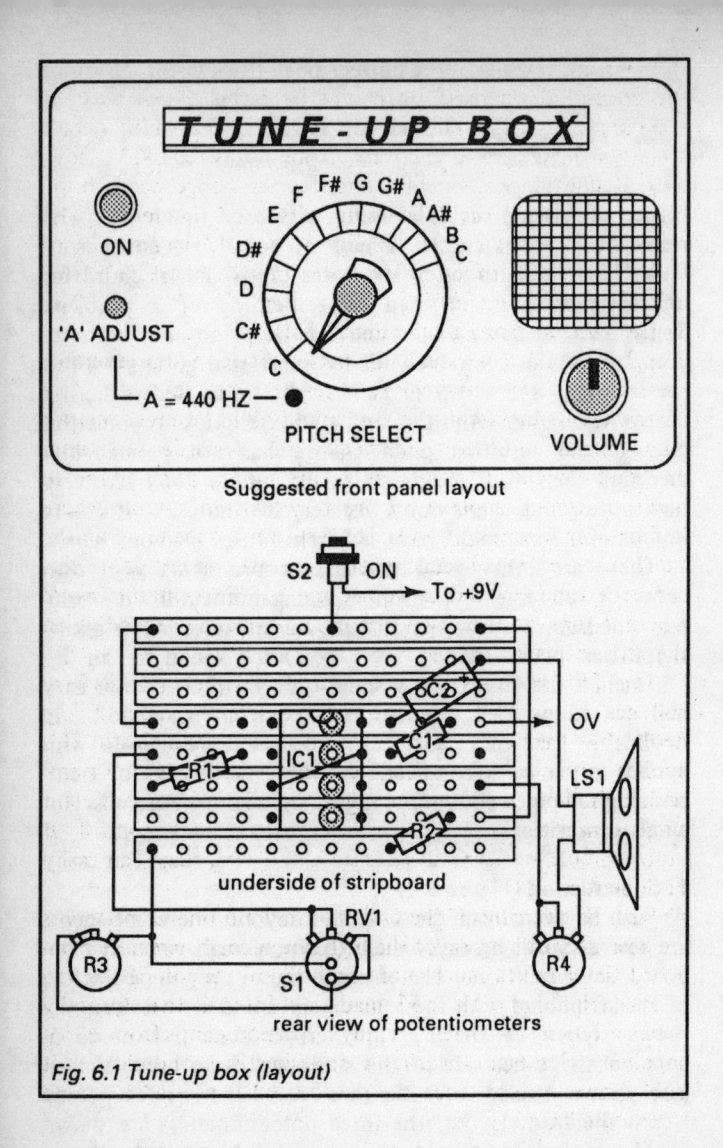

If these are reversed, the conventional clockwise rotation of the control to increase pitch will be reversed and also the scale markings. The value chosen for RV1 covers a full octave scale which, in any key, covers all the notes required. However, if due to component tolerances this range needs to be higher or lower, then the value of C1 or the series fixed resistor R2 can be changed slightly to suit. A larger value of C or R lowers the pitch of the note, and the converse is true.

2 The Tune-up Box Circuit (Figure 6.2)

The Tune-up Box is essentially a pocket-size, audio generator for selecting and sounding the notes of a musical scale. Any of the oscillator circuits in this book could be used, but for compactness and wiring simplicity the versatile 555 timer integrated circuit is suggested. This chip is popular for all kinds of timing applications in its *monostable mode* where it flips over to one of two states for a time depending on the value of a resistance-capacity (RC) circuit. However, in this case, for tone generation we use the 555 in its *astable mode* where it flips between two states at a frequency depending on the value of its RC circuit.

The 555 provides a stable output frequency. Battery voltages from 9.5V down to 7V were tested without any detectable frequency drift. This makes it suitable for this application and others where frequency stability is important. Although there is no output stage, the circuit gives sufficient audio power to warrant the inclusion of a volume control. If more volume is required an extra amplifying stage can easily be added.

With this version of the circuit, the drain on the battery is less than 10mA because of the high resistance between IC1 pin 7 and the + rail (a much more economical version of the 555 is available, the ICM 7555 made by Intersil, that typically draws a quiescent current of only 100 micro-amps from a 15V supply). However, the circuit described is economical as it only draws current when the pushbutton is pressed to sound the note selected.

The frequency-determining components are capacitor C1 and the network of resistors that connect it to the positive supply rail. In each operating cycle, C1 charges via this resistor

Fig. 6.2 Tune-up box (circuit)

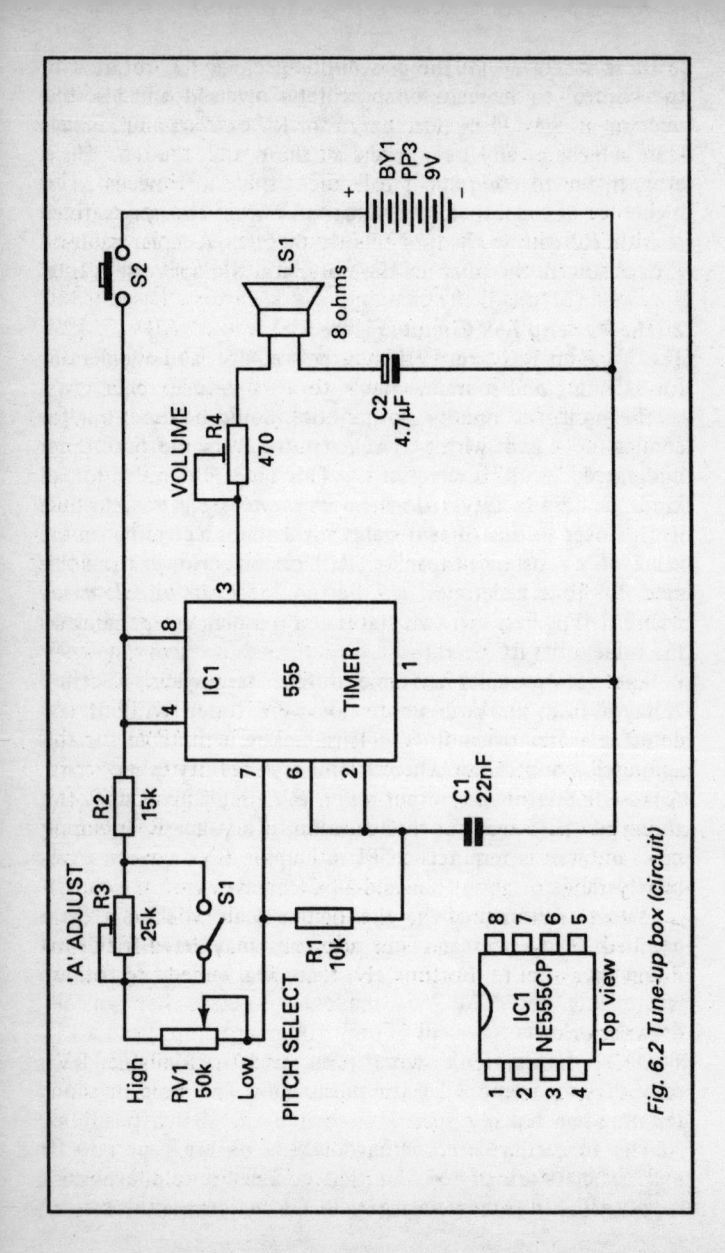

chain from the + rail, and is discharged via R1 only. The waveshape, i.e. the mark-to-space ratio of each period of the rectangular waveform will vary over the frequency range. This is because R1 is a fixed resistor and the resistance between the + rail and pin 7 is variable. However, the difference in tonal quality does not matter in this application.

Although the basic circuit is simple, some explanation of the resistor network is necessary. With RV1 set fully anti-clockwise (S1 open) the charging resistance for C1 is at maximum. It consists of R2, preset R3, the whole of RV1's 50k, and R1. This fully anti-clockwise position of RV1 sounds the lowest note, which is adjustable to a suitable frequency by the only variable resistor now in circuit, R3. This preset is labelled 'A' Adjust because the violins of an orchestra tune-up to the note 'A' (440 vibrations per second). In this way, the fixed standard facility gives the user a note that is easy to find on the night by just pressing the pushbutton S2. Other notes can be fixed tuned instead of 'A' For instance, in the brass band world, a B-flat and an E-flat are more useful references than an 'A'. For the other notes in a scale, the front-panel 'Pitch Select' variable resistor RV1 turned clockwise, and R3 is therefore short-circuited by S1. The charging resistance for C1 is now R2, the variable resistance RV1, and R1. With the aid of a piano or another musical instrument, RV1 can be calibrated to indicate a complete scale of notes or more. Because R3 is not in circuit when RV1 is switched on, the tuning of the fixed standard does not affect the RV1 tuning scale and vice versa. The 50k value for RV1 covers a frequency range of about one-and-a-half octaves with C1 chosen as 22nF. Unfortunately, the octave scale will be a little cramped at the top end and accuracy may be difficult to obtain as a result. Fortunately, there is a remedy to follow.

3 Variations

If a 12-position rotary switch (and space) is available, RV1 can be replaced by 12 preset resistors. The scale divisions will then be equally spaced to match the switch positions. This is obviously an advantage where accuracy of note is important. Alternatively, the notes of a scale could be obtained by using a rotary switch to bring in different values of capacitance for C1. These would have to be selected on test as tolerances vary so much. On test, with a 22nF for C1, with 65k between IC1 pin 7 and the + rail, a 'B♭' sounded. For the 'B♭' one octave lower, C1 needs to be about 47nF — just over twice the value of capacitor per octave drop. So to cover an octave from B♭ to B♭, twelve capacitor values would be needed between 22nF and 47nF. If a note sounds slightly higher than it should, it means the value of the tuning capacitor selected is too small. Remember that unlike resistors, capacitors in parallel add up, so a much smaller value of capacitor can be connected in parallel to 'pad' such a tuning capacitor to bring it down into tune. Overall tuning can be done by RV1, which could be a preset resistor in place of R3.

Components, Tune-up Box

Resistors
R1	10k
R2	15k
R3	25k preset
R4	470 ohms preset

Potentiometer
RV1	50k potentiometer with switch (Pitch Select)

Capacitors
C1	22nF
C2	4.7μF 10V elect

Semiconductors
IC1	NE555CP Timer

Loudspeaker
LS1	8 ohms

Switches
S1	S.P.S.T. (on RV1)
S2	Push to make

Miscellaneous
Case, stripboard, 9 volt battery (BY1), connector to suit, wire, etc.

Chapter 7

MELODY RANGER

Melodies are usually restricted to a range of two octaves, for which those of us with limited voice ranges are truly thankful. Here is a two-octave stylus organ with a piano-type keyboard, and as no self-respecting melody goes above F or G, it conveniently covers the range $G_1 - G - G^1$. Although this Melody Ranger is monophonic, i.e. plays only one note at a time, a vibrato circuit can be switched in as desired to make the sound more interesting.

1 Vibrato

Vibrato is that pleasant warming effect that violinists get with a slow shake of the left hand – or is that Boy Scouts? Strictly speaking, vibrato is a slow variation in pitch, whereas a similar effect, tremolo is a slow variation in intensity. If you have made up one of the earlier circuits, for example, the Electronic Sol-Fa or the Rhythm Checker, you can hear the effect of vibrato for yourself. Sound out a note and notice how bare and unmusical it is. Now try waving your hand backwards and forwards across the front of the loudspeaker at a speed of about six vibrations per second. Notice how much more musical this sounds. Other methods are used to obtain vibrato besides controlled vibrations of fingers on strings. Brass players shake fingers on valves, shake their heads and lips, and trombonists have the added luxury of slide oscillation to produce vibrato. As a trombonist, I find that the depth of the vibration should not exceed more than about a quarter of a tone otherwise the effect sounds ugly. Normally, it should be used sparingly. Electrically, vibrato is produced in some electronic organs by rotating loudspeakers, and in vibraphones by rotating paddles in the ends of the resonators. Electronically, vibrato is produced by an oscillator running at about 5 to 8 Hz.

Fig. 7.1 Melody Ranger (keyboard layout)

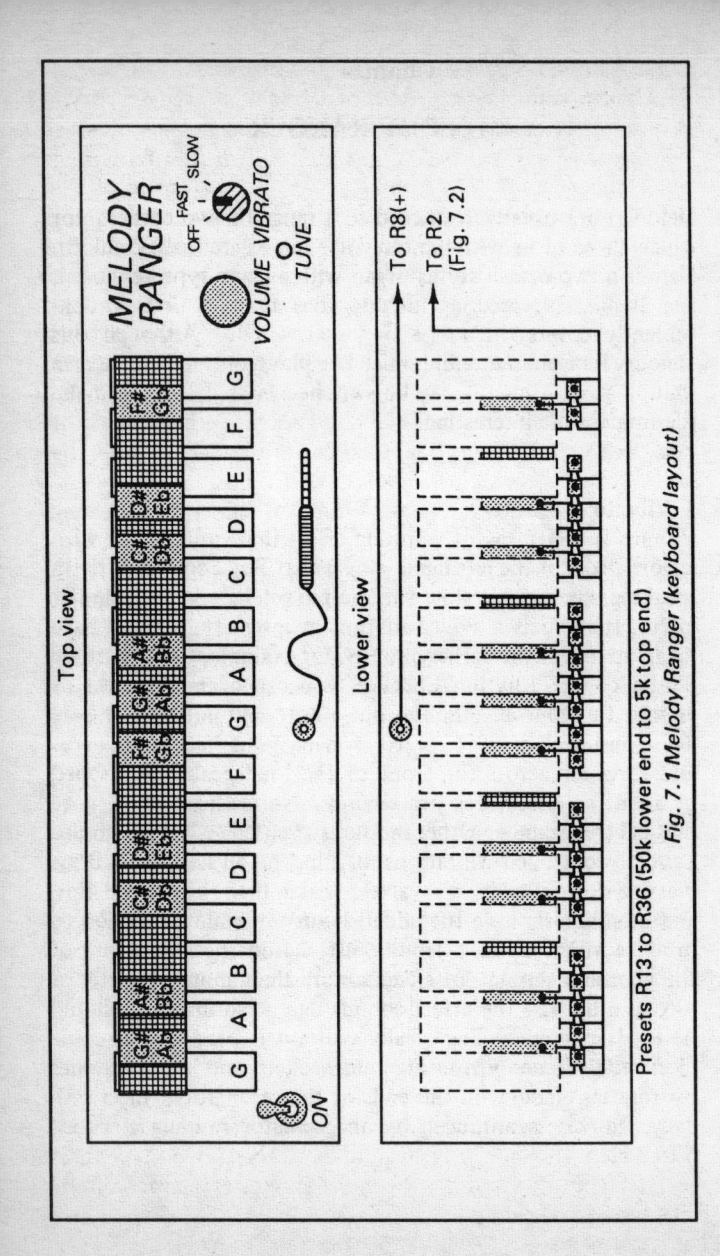

Presets R13 to R36 (50k lower end to 5k top end)

2 Mechanical Construction (Figure 7.1)

The keyboard dictates the size of the Melody Ranger. If you are good at mechanical work, you could reduce the size considerably. However, the example shown uses fairly large key-plates (6cm × 1.5cm, and 3cm × 1.5cm) made of thin aluminium or stainless-steel plate. The 25 key-plates required for two octaves will comfortably fit on a paxolin, Formica or 3-ply front panel 40cm × 10cm. The 0.5cm spacing between the larger key-plates (the diatonic notes) is left for two reasons. It prevents the stylus short-circuiting two adjacent keys which would result in a rogue note sounding as two tuning preset resistors would be connected in parallel; also the space is needed to drill the holes for the fixing bolts of the ten shorter key-plates (chromatic notes). Make sure that these bolts are clear of the two key-plates either side.

The key-mounting details are given in Figure 7.3. The main purpose of the smaller panel is to insulate the smaller key-plates from the larger key-plates that lie underneath. No doubt constructors will think of other ways of making a dummy keyboard. One was tried using copper-clad board. If this is used, there is no need for the smaller panel. The only problem is that the copper cladding tends to tarnish quickly. Whatever materials are used, the important thing is to ensure that all the key-plates are insulated from each other, are clean (make good low-resistance contact with the stylus), and are in a logical sequence for playing.

A suitable box to house the circuit components can be made or acquired for the front panel to fit on. It need not be too deep as it houses only a circuit board, a 9V battery and the loudspeaker. If the loudspeaker is mounted on the bottom of the box, fit four rubber feet at the box corners to allow clearance, otherwise the sound will be muffled.

3 Circuit (Figure 7.2)

Integrated circuit chips (555s for example) could have been used for this project, but four discrete transistors were chosen, mainly to give a better idea of how the three sub-circuits work. You will recognise the multivibrator oscillator from the electronic sol-fa circuit.

Fig. 7.2 Melody Ranger (circuit)

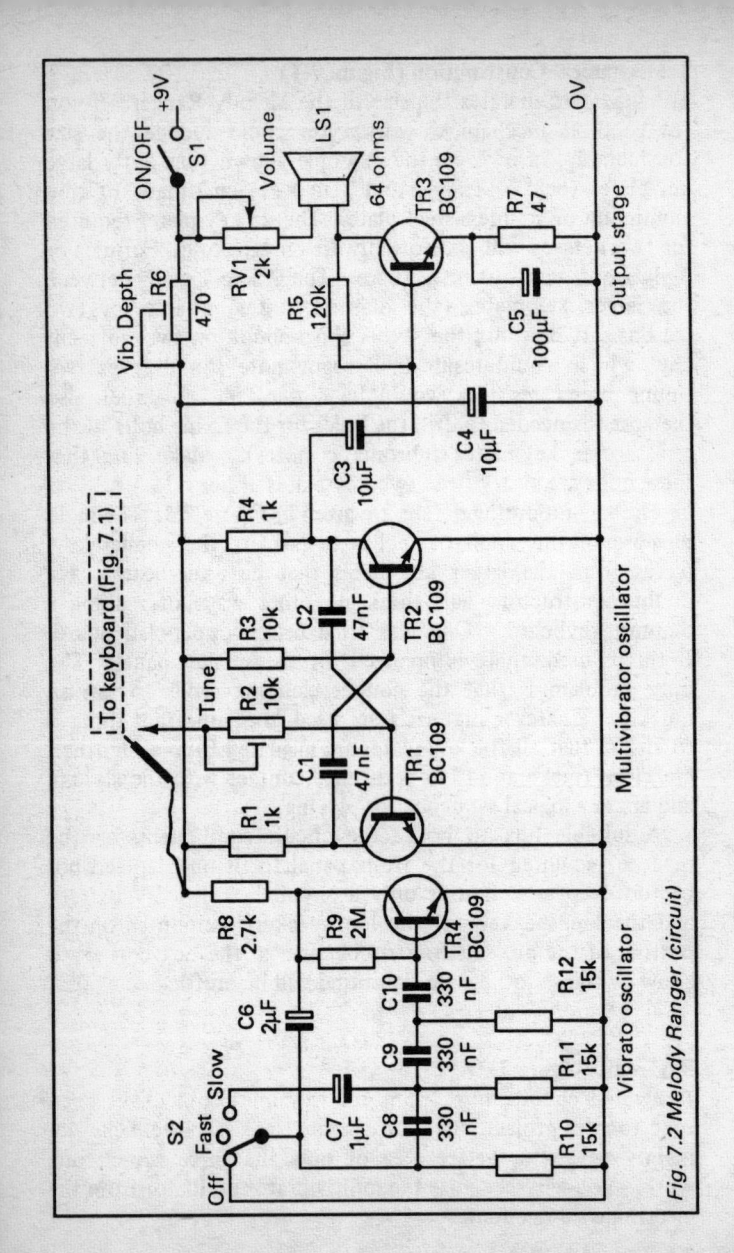

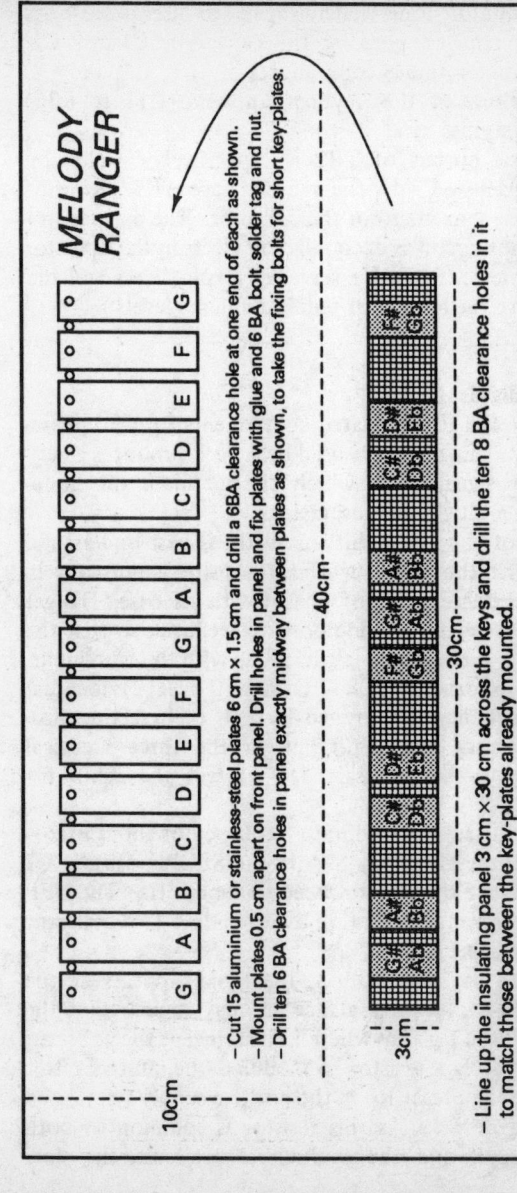

MELODY RANGER

- Cut 15 aluminium or stainless-steel plates 6 cm × 1.5 cm and drill a 6BA clearance hole at one end of each as shown.
- Mount plates 0.5 cm apart on front panel. Drill holes in panel and fix plates with glue and 6 BA bolt, solder tag and nut.
- Drill ten 6 BA clearance holes in panel exactly midway between plates as shown, to take the fixing bolts for short key-plates.

- Line up the insulating panel 3 cm × 30 cm across the keys and drill the ten 8 BA clearance holes in it to match those between the key-plates already mounted.
- Cut and drill ten key-plates 3 cm × 1.5 cm and mount them on smaller panel. Line up holes using the 6 BA bolts and glue key-plates to smaller panel.
- Fit panel over larger panel, align the ten bolts in the mounting holes and secure underneath with solder tags and nuts.
- One leg of each of the 24 presets connects to a solder tag; an adjacent leg is soldered to the common + rail (see Fig. 7.1).

Fig. 7.3 Melody Ranger (key mounting details)

The multivibrator tone generating stage uses transistors TR1 and TR2 cross-coupled by the capacitors C1 and C2. The frequency-determining components are C1, R3, and C2, preset R2 and one of the keyboard presets (R13 to R36) selected by the stylus.

The collector output of TR2 is capacitively-coupled by electrolytic capacitor C3 to the output stage TR3. Resistor R5 provides base current from the collector. The bias resistor R7 in the emitter circuit is decoupled by electrolytic capacitor C5. The variable resistor RV1 serves as a volume control and helps to conserve the battery at reduced volume levels.

4 Vibrato Oscillator

Transistor TR4 and its associated components forms a phase-shift oscillator. This circuit is used because it produces a very good sinewave signal with which to modulate our note-producer — the multivibrator oscillator.

The action of the phase-shift oscillator is best understood if we realise that the collector of a transistor is normally in anti-phase (180 degrees out of phase) with its base. To get oscillations we need to provide positive feedback so that the collector output is coupled back in phase with the base input. The 3-section CR network (C8 — C10, R10 — R12) does just that. At a frequency determined by CR, each section provides a 60 degrees phase-shift. Over the three identical sections, this gives the necessary 180 degrees phase-shift for oscillation.

The values chosen give a vibrato frequency of about 8Hz — the FAST position of switch S2. In the SLOW position, C7 shunts C8 to reduce the vibrato speed to about 6Hz. The OFF position of S2 prevents oscillation by grounding TR4 collector via the coupling capacitor C6.

When vibrato is switched on, the signal appears at the collector of TR4. At first glance, it may appear that the vibrato signal doesn't go anywhere. The signal at the collector could be coupled via a resistor to modulate the multivibrator, but in this case the preset R6 in the positive supply rail is used as the coupling device. As this resistor is common to both the vibrato circuit and the multivibrator circuit, the slow variation in current through it caused by the vibrato circuit modulates the multivibrator. Varying the value of R6 varies the depth of modulation. This should be set before tuning the instrument. The Tune preset R2 should be set with the stylus touching the top G; i.e. the note without a tuning preset. Tuning is similar to that described for the Electronic Sol-fa unit.

Over the volume range, the current consumed by the circuit varies from about 10mA to 30mA, for a supply voltage of 9V. If desired, a low-impedance speaker can be used in place of the high-impedance speaker, fed from the collector of TR3 via a step-down output transformer.

5 Circuit Construction

The stripboard layout for the multivibrator stage is similar to that described for the Electronic Sol-fa project. Again, the positioning of components is not critical. If desired, the multivibrator and output stage can be made up on a separate piece of stripboard and proved before tackling the vibrato stage.

A suggested layout for the vibrato oscillator is given in Figure 7.4. For simplicity, all components and connections are shown on the track-side, but the choice is yours. Don't forget to make the four breaks in the stripboard as indicated! As the vibrato stage is quite independent, this could be used to add to some of the other projects.

Components, Melody Ranger

Resistors

R1	1k
R2	10k 0.1 watt preset
R3	10k
R4	1k
R5	120k
R6	470 ohms 0.1 watt preset
R7	47 ohms
R8	2.7k

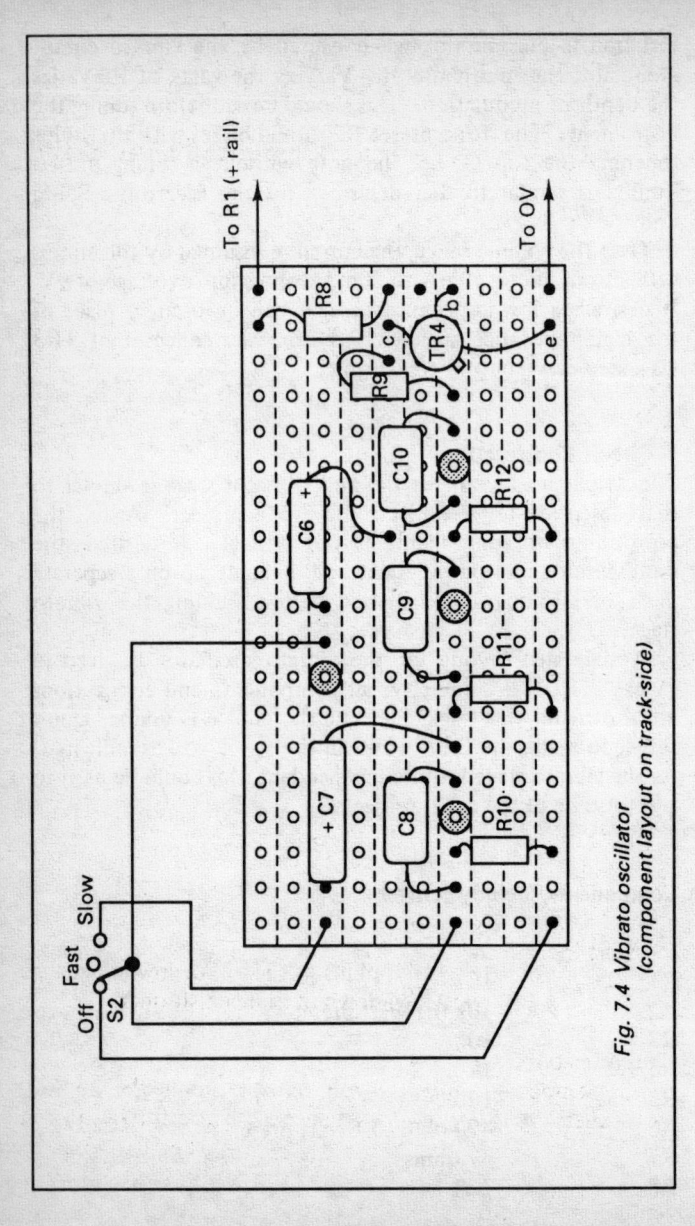

Fig. 7.4 Vibrato oscillator (component layout on track-side)

R9	2M
R10	15k
R11	15k
R12	15k
R13 to R36	50k lower to 5k top end, 0.1 watt presets

Potentiometer
RV1 2k (Volume)

Capacitors
C1	47nF plastic foil
C2	47nF plastic foil
C3	10µF 10V
C4	10µF 10V
C5	100µF 10V
C6	2µF 10V
C7	1µF 10V
C8	330nF plastic foil
C9	330nF plastic foil
C10	330nF plastic foil

Semiconductors
TR1	BC109
TR2	BC109
TR3	BC109
TR4	BC109

Switches
S1	S.P.S.T.
S2	S.P. 3-way

Loudspeaker
LS1 25 to 80 ohms as shown, or low-impedance with step-down output transformer.

Miscellaneous
Case, component board, 9-volt battery, connector to suit, wire, metal strip, panel or metallic-clad board for key-plates.

Chapter 8

APPEALING HANDBELLS

Bell effects are not easy to simulate electronically as bell waveforms are rather complex. Although these 'appealing' electronic handbells are in no way intended as serious contenders to the traditional handbells, they provide an entertaining novelty item for a musical programme. You will of course need about eight of them if your team is going to ring the changes. However, the handbell circuits are simple and much cheaper than the real thing.

The biggest problem was how to make a handful of electronic components look like a handbell, or sets of handbells of different sizes. Bell-shaped project boxes are not yet listed in manufacturers' catalogues so it was necessary to improvise. The prototype circuits were housed in different sized plastic flowerpots and sprayed with gold paint, courtesy of an auto aerosol can. Alternatively, a tin of gold paint sold at most craft and model shops will add the required lustre.

1 Mechanical Construction (Figure 8.1)

A length of three-quarter-inch diameter plastic tubing from the plumbing section of a DIY store will provide all the handles you need. Saw off a 5-inch length for each handle. Drill a hole 2 inches from one end to take the pushbutton switch. You will also need to drill a slightly larger hole on the opposite side of the tube to feed in the switch. Try the switch for size, but do not fit it at this stage. Now make about eight equidistant saw-cuts of 1-inch from this end of the tube and gently splay the sections out as shown.

Cut a three-quarter-inch central hole in the base of the flowerpot. Feed the handle through this ready for gluing. The splayed out sections will provide more gluing area, and help to strengthen the plastic flowerpot base which may be a little flimsy. Before gluing, solder an eight-inch length of insulated wire to each of the switch contacts. Mount the switch through the larger hole and feed the switch wires down the tube into the bell section ready for connection. Now glue

Fig. 8.1 Handbell construction

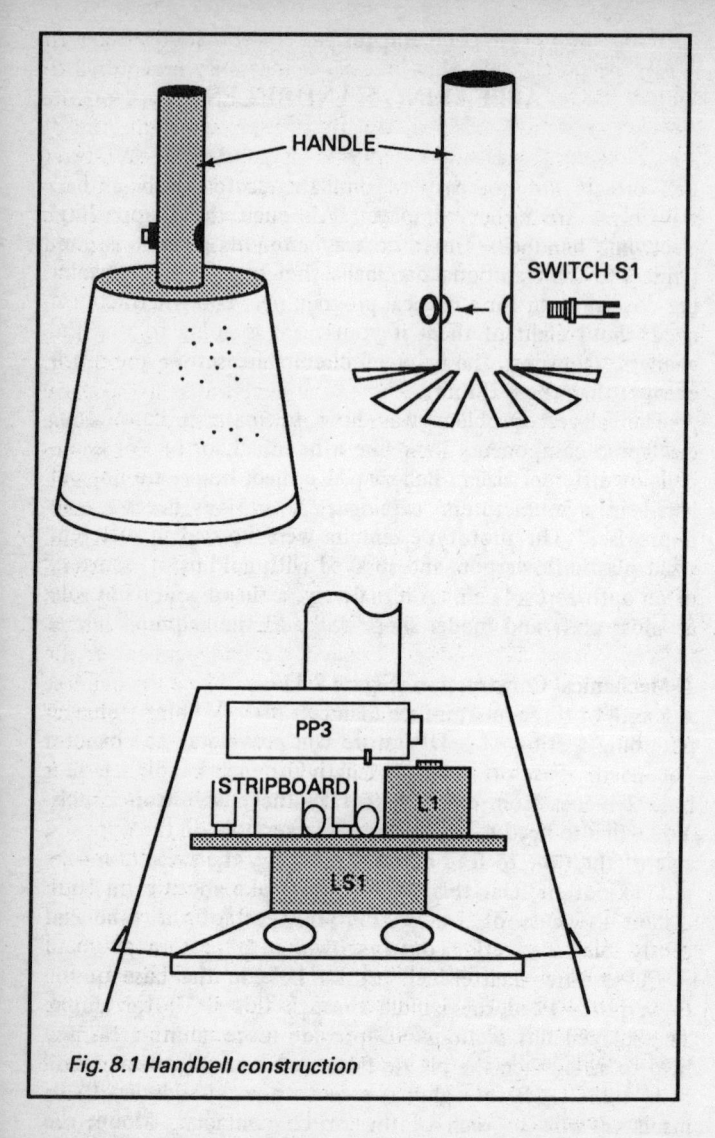

the handle to the bell. Make sure it stands vertical and allow to dry.

The few circuit components, battery and loudspeaker fit easily inside the bell area, but some ingenuity is required to mount them. One way is to mount the stripboard on the speaker, suitably insulated, and fix the speaker to the rim of the flowerpot with wire supports as suggested. Whatever method is used, remember that the battery needs to be accessible. And, as for effect it is necessary to shake the handbell, make sure that all the components are well secured or you could literally drop a clanger!

2 The Circuit (Figure 8.2)

Each circuit uses a Hartley oscillator as the tone generator. These were firm favourites for tone generators in the early electronic organs. The positive feedback necessary to produce oscillation is obtained through the coil L1, the centre-tap going via R3 to the emitter of TR1. The frequency is dependent on the value of inductance of L1 and the value of capacitance (C2) across it. Typical values for C2 over about an octave are 200nF (low notes) to 47nF (high notes). The coil is wound on the former of a Ferroxcube pot core. Typical coils consist of 500 turns of 39-gauge enamelled wire, centre-tapped. With this coil, the frequency is independent of the collector voltage. Use is made of this to produce the bell-like decay. When pushbutton S1 is pressed, the 9V battery charges up the electrolytic capacitor C3 and provides the collector voltage to set the oscillator circuit TR1 in motion. When S1 is released, C3 provides the collector voltage to sustain oscillation until the capacitor C3 gradually discharges.

This gives a chime-like decay, the time of which can be altered by changing the value of C3. The value of $10\mu F$ gives a decay of about one second. Increasing the value of C3 lengthens the decay time. It is interesting to experiment with the value of C3. A $470\mu F$ electrolytic maintained oscillation for about half a minute. For handbells, however, one or two seconds seems appropriate. Although this Hartley circuit will oscillate using a centre-tapped laminated-core coil for L1, the note tends to drift in frequency as it decays. Using a Ferroxcube core for L1 obviates this problem. The prototype coils were wound by hand (a little tedious, but less than a quarter of an hour per coil!). When the centre-tap point is

Fig. 8.2 Handbell (circuit)

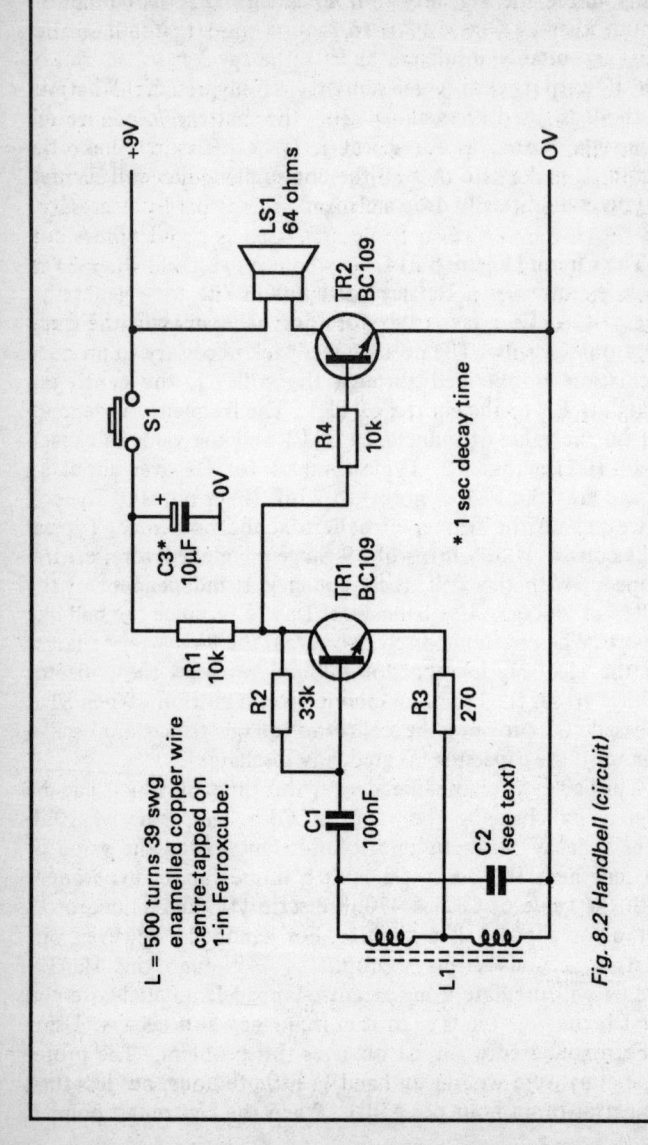

reached, bring out a loop for soldering purposes. Be sure to continue winding in the same direction for the second half of the coil otherwise your halves will be wound in antiphase and the circuit will not oscillate.

Note that there is no ON/OFF switch in circuit, the output transistor and loudspeaker being permanently connected across the 9-volt supply. An extra ON/OFF switch could be incorporated if there is room, but as the quiescent current measured was a mere 10 micro-amps this is hardly necessary. The initial current when S1 is depressed is about 60mA but drops to its quiescent state after about one second when S1 is released. Flowerpot power — about 200mW!

A stripboard layout for the handbell is shown in Figure 8.3.

Components, Appealing Handbells

Resistors
R1   10k
R2   33k
R3   270 ohms
R4   10k

Capacitors
C1   100nF plastic foil
C2   47nF to 200nF (see text)
C3   10μF 10V (or larger, see text)

Semiconductors
TR1   BC109
TR2   BC109

Switches
S1   push to make

Loudspeaker
LS1   miniature 64 ohms

Miscellaneous
Plastic tubing, plastic flowerpot, 9-volt battery (PP3) and connector, stripboard, wire.

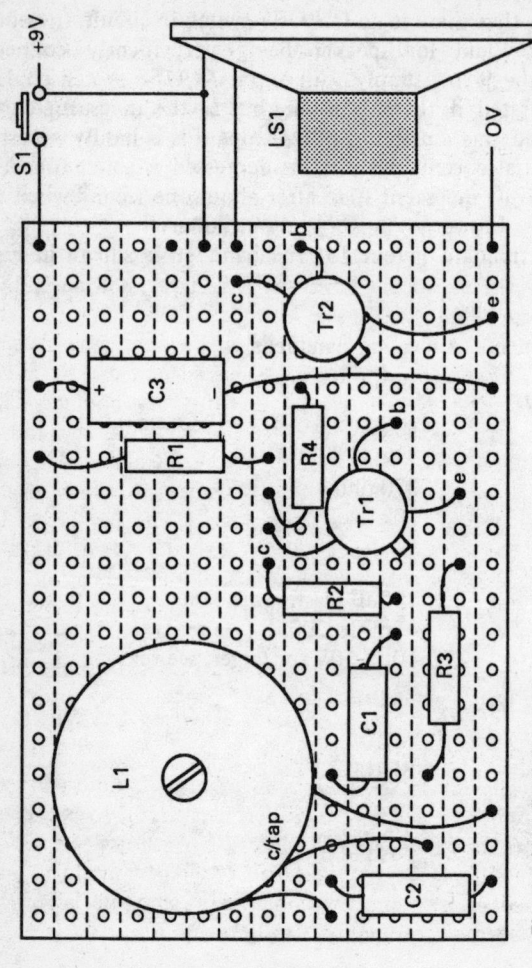

Fig. 8.3 Handbell (layout)

Chapter 9

THE ELEXYLOPHONE

The Elexylophone or electronic xylophone can be simple or as sophisticated as interest, time and ingenuity allows. For this reason, it is suggested that a modular, building block method is adopted as you may wish to extend or experiment later. For instance, your xylophone may be restricted to one octave of the diatonic notes of a major scale; e.g. C, D, E, F, G, A, B, C. In this case you will need to build only eight oscillator circuits. However, if you want to include all the chromatic notes (accidentals) then you will need thirteen oscillators as shown in the example. The dummy keyboard (Figure 9.1) is simple to adapt to whatever scheme is required. It can have as many notes as desired and can be any size. Unlike the traditional xylophone, small can be loud, depending on the output amplifier. But more about this and other variations after we have introduced the theme.

The well-tried Hartley circuits, used for the electronic handbells form the tone generators. The circuit is simple, stable, has a large output and can easily be tuned. It produces bright, resonant sounds when shocked into oscillation by a tap from a beater on the key-plates. Since the characteristic xylophone sound has a very short decay time, notice that the electrolytic capacitors (C3) and collector resistors are smaller than for the handbells.

1 The Circuit Modules (Figure 9.2)
Although, for clarity, the circuit of one of the oscillators and the amplifier are shown together, it is useful to build these as separate modules on small pieces of stripboard; i.e. an amplifier module to which several oscillator modules can be attached. There are few components and the largest layout board need not exceed 2-in square.

The oscillator
When Hartley devised this LC (inductance-capacitance) oscillator back in the 'Valve Age', transistors were not even a

Fig. 9.1 Elexylophone (keyboard layout)

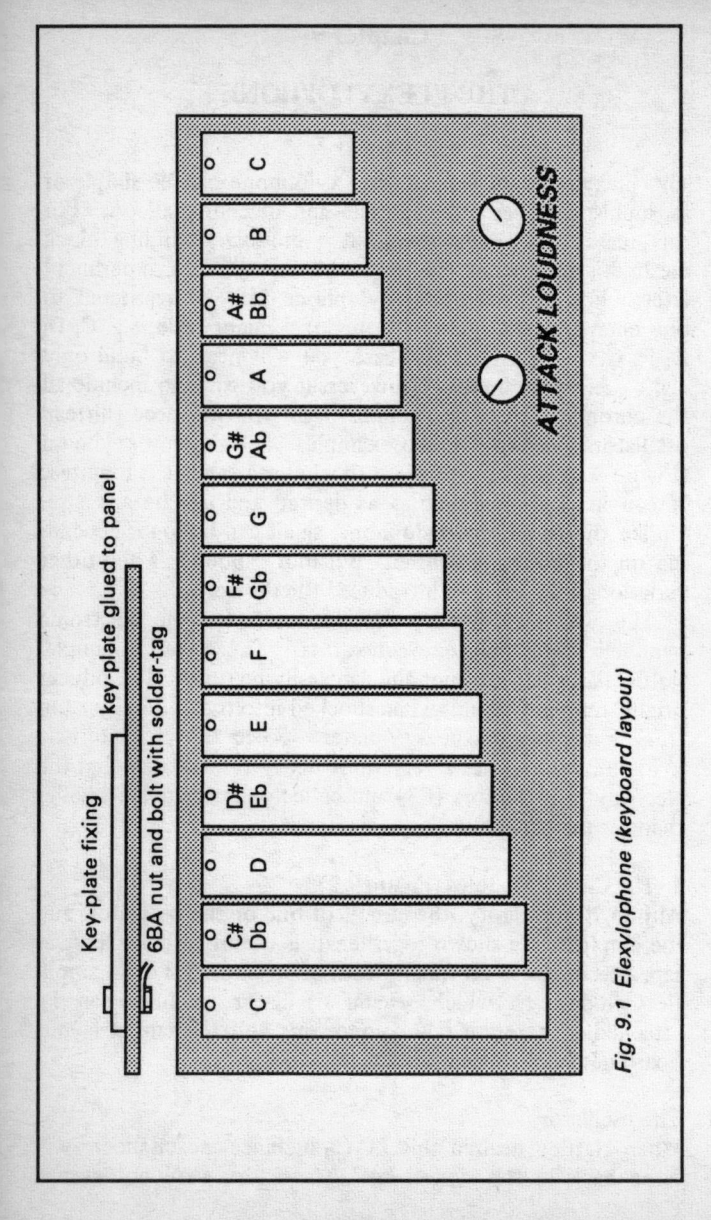

Fig. 9.2 Elexylophone (circuit)

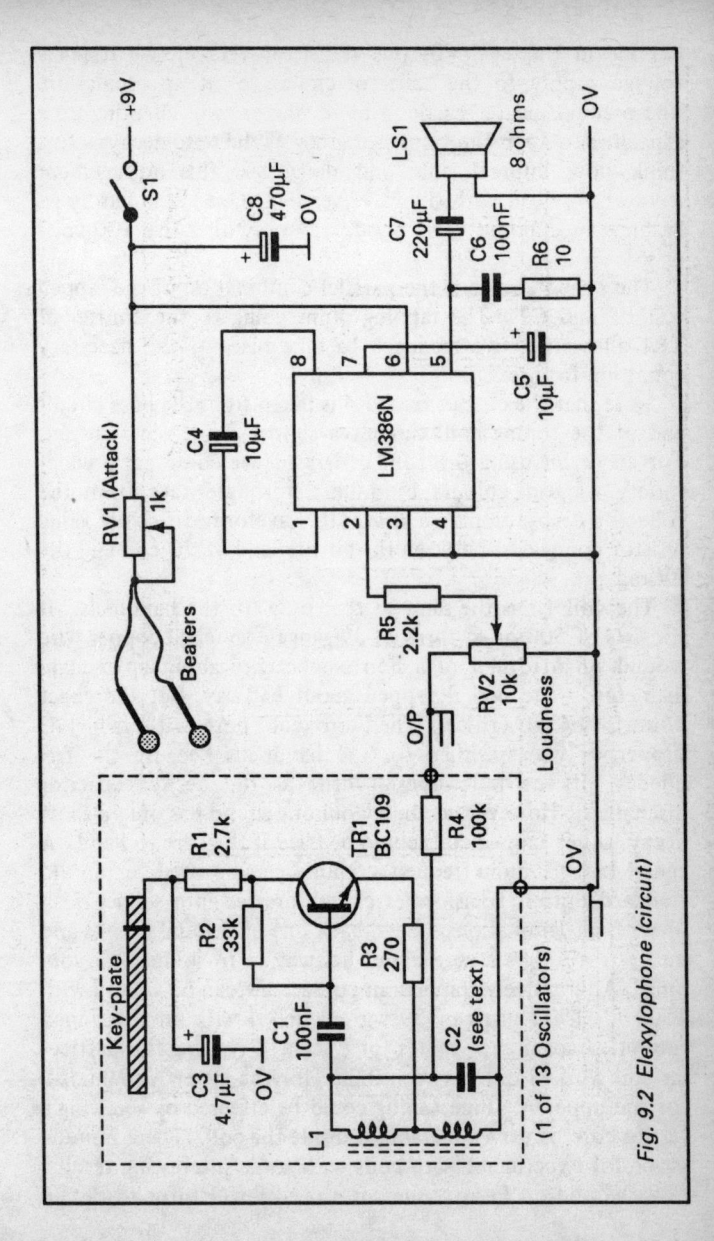

twinkle in the eye. In this transistor version, we touch a voltage supply to the collector circuit to set up oscillation. The metallic beater carries a mere volt or two, charging up a capacitor to keep the transistor going as the note decays. Just think how impracticable and dangerous this arrangement would have been with the valve version. One could hardly go bashing an exposed valve anode around with a live 120 volts or so!

The tuned circuit is the parallel combination of the tapped coil L1 and C2. The tapping point going to the emitter of TR1 allows positive feedback to take place — the necessary condition for oscillation.

Note that the output from R4 is taken from the base circuit end of the tuning coil; this gives almost a sine-wave output. For an organ using this tone generator, we could get a wider variety of tone-colours by taking a square-wave from the collector or by tapping a sawtooth waveform off a low value resistor connected between the bottom end of the coil and the 0V rail.

The coil L1 is the same as that used for the handbells. It consists of 500 or so turns of 39 gauge enamelled copper wire wound on a former of a Ferroxcube core about an inch in diameter. The coil is tapped about halfway, but the exact point is not too critical. The Ferroxcube core, with its high-Q properties was specified for the handbells because the frequency of the note doesn't drift as the decay capacitor discharges. However, as the xylophone sound has only a short decay, other tapped coils could be tried if they are to hand. A centre-tapped audio-frequency choke can be used, or an old standard output transformer can be pressed into service with or without laminations. Bend back one of the end cheeks and pull out a loop of wire about halfway in to solder on a tapping. Alternatively, tapped air-cored coils can be wound with 1500–0–1500 turns of 39swg enamelled wire on a half-inch former. Capacitance values for a scale of A using this particular coil worked out between $0.2\mu F$ for the lower A to $0.05\mu F$ for the upper A. Fine tuning could be effected by screwing a ferrite core or even a metal bolt inside the coil. There is much scope for experiment with coils with some interesting results. Coils wound on Ferroxcube cores need fewer turns to obtain the required inductance. For middle C, a coil inductance of about 200mH is needed with a 0.5μF tuning capacitor.

Depending on circuit parameters, the value of electrolytic capacitor C3 may need to be finally adjusted to give the correct decay time. Also, R4 could be selected on test, or a variable resistance, to adjust individual loudness levels to compensate for room acoustics or speaker characteristics. Increase the value of R4 to reduce the input level to the amplifier control RV2.

The purpose of the variable resistor RV1 in the +ve line to the beaters is to vary the attack of the note. With no resistance in circuit, the initial attack to the note will be rather brittle, like a musical sforzando (>) note. Adding more resistance gives a mellower, more subdued start to the sound.

A stripboard layout for the oscillator is shown in Figure 9.3.

The amplifier

The widely-available LM386N audio amplifier IC is used as the common output stage. The input is fed from the slider of the loudness control RV2 via resistor R5. Pin connections for the LM386N are shown on the circuit diagram in their correct location viewed from the top-side of the IC. Make sure that the polarities shown for the electrolytic capacitors are observed.

The oscillators provide more than adequate output. Typically, a level into the amplifier of 500mV pk-pk will give an output of about 250mW r.m.s. into 8 ohms. Loudspeakers with higher impedances, for instance 64 ohms, can also be used. A typical layout for the amplifier stage is given in Figure 9.4. This self-contained amplifier unit can be usefully employed in other projects.

2 Construction

A complete octave unit can easily be mounted in a shallow box about 3-in deep under the keyboard panel shown in Figure 9.1. If a small speaker is used, there is adequate room for this on the front panel at the upper end of the keyboard. Each oscillator module suggested needs a surface area of about 2in × 2in, and these can be mounted in the base of the box.

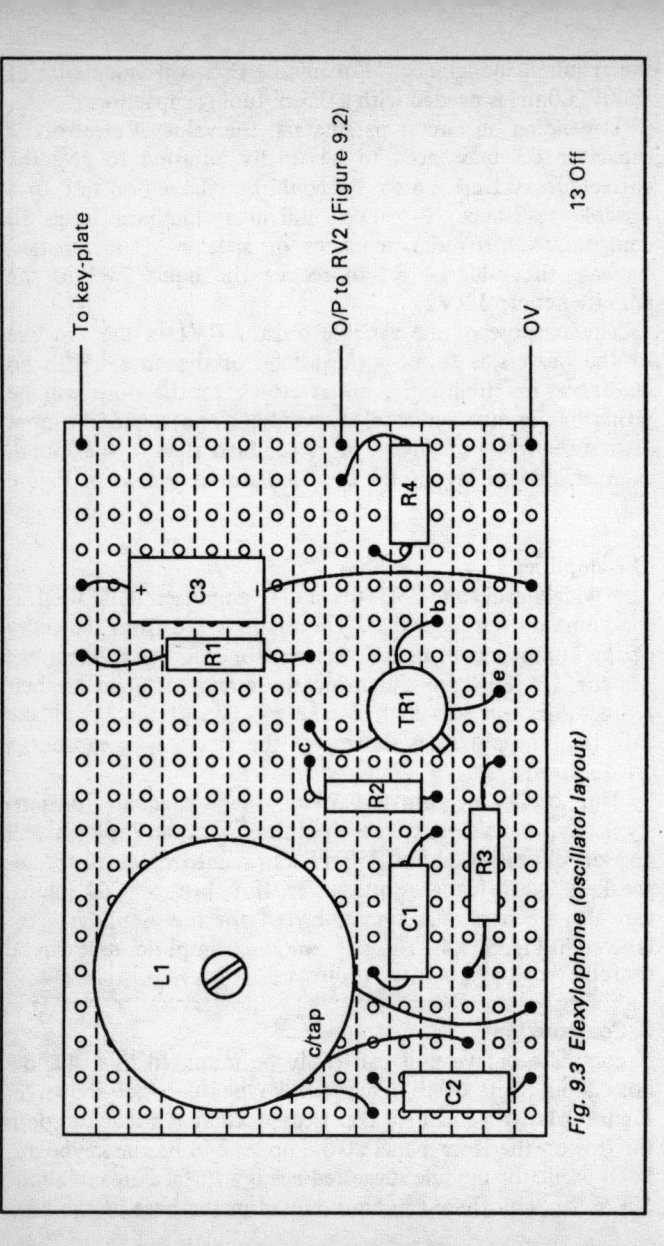

Fig. 9.3 Elexylophone (oscillator layout)

Fig. 9.4 Elexylophone (amplifier layout)

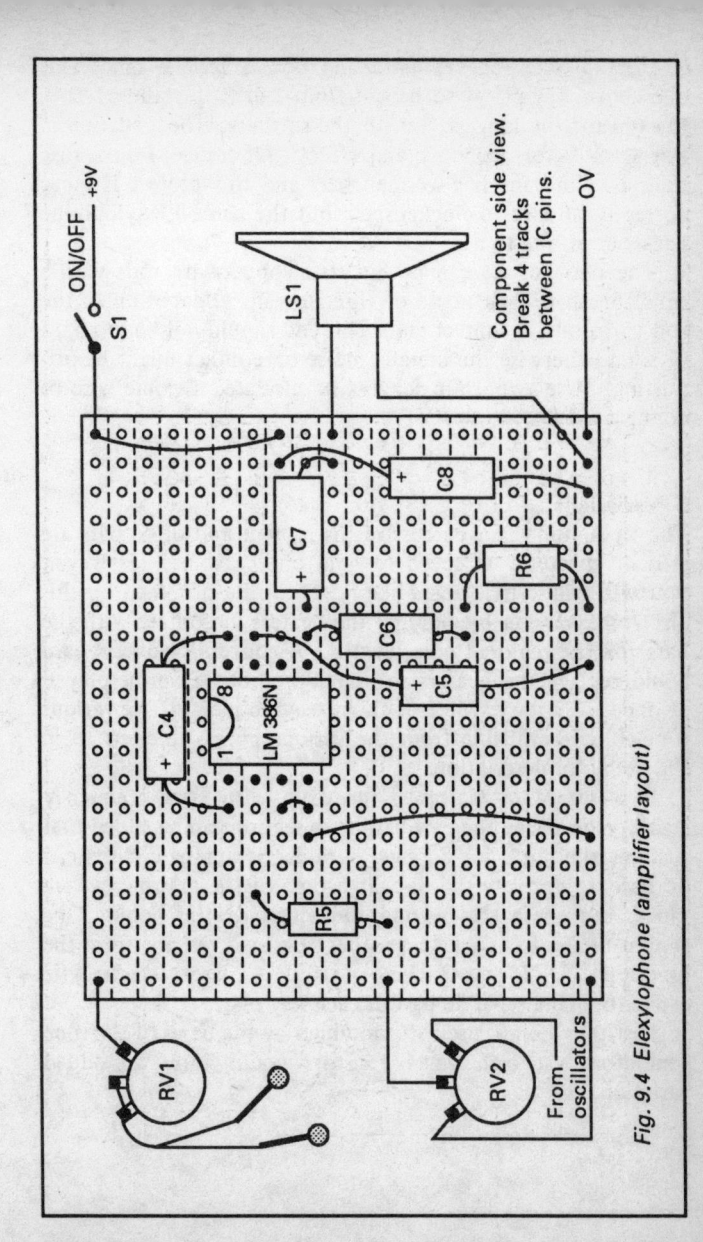

A suitable size for keyboard and box is 16in × 8in. This allows for key-plates to be cut from 1-in strip stainless-steel, say 6in for the longest, 3in for the smallest. The gradation in size is only for economy and effect. Of course, I do realise that a xylophone has wooden bars and this project is more correctly an electro-glockenspeil, but the name Elexylophone has a better *ring* to it!

The two beaters can be 8in stiff copper-wire rods with a small brush or ball made of thinner wire soldered on to the rod to form the contact end. This end should not be too hard or solid otherwise the metallic noise on contact might be off-putting. Use two 18-in lengths of insulated flexible wire to connect the beaters to RV1.

3 Variations

The xylophone is a percussive instrument and often bars are struck repeatedly at speed to hold on the sound. However, you will realise that organ-like notes can be sustained on this electronic version by holding the beaters in contact with the bars for the required note lengths. A conventional keyboard could replace the beaters so that several notes can be played at once. A rotary switch could be used to pick off the various tone-colours available from the various points referred to in the oscillator description.

If a set of oscillators is made up with good frequency stability on decay then a vibraphone section can be added that sustains the notes for a longer period. Switching circuits can be used to regulate the discharge time of the existing capacitance, but these are beyond the scope of this book. One simple method is to take another 10k load resistor from the junction of R1 and TR1 to a $10\mu F$ or larger electrolytic capacitor and a separate part of each key-plate.

Lastly, a vibrato oscillator modulating the bases of the tone generators via high value resistors could offer an added variation.

Components, Elexylophone

Per Oscillator

Resistors
R1　　　　　2.7k
R2　　　　　33k
R3　　　　　270
R4　　　　　100k (see text)

Capacitors
C1　　　　　100nF plastic foil
C2　　　　　47nF to 500nF (see text)
C3　　　　　4.7μF 10V

Coils
L1　　　　　(see text)

Semiconductors
TR1　　　　 BC109

Amplifier

Resistors
R5　　　　　2.2k
R6　　　　　10 ohms

Potentiometers
RV1　　　　 1k
RV2　　　　 10k

Capacitors
C4　　　　　10μF 10V
C5　　　　　10μF 10V
C6　　　　　100nF plastic foil
C7　　　　　220μF 10V
C8　　　　　470μF 10V

Semiconductors
IC1　　　　 LM386N

Loudspeaker
LS1			8 ohms

Switches
S1			S.P.S.T.

Miscellaneous
1-in strip stainless-steel for key-plates, suitable panel and case, stripboard, 9-volt battery, beaters, wire, 6BA nuts, bolts, solder-tags.

Chapter 10

THE RHYTHM SETTER

An Audio/visual Metronome for all Times!

1 Introduction

Metronomes keep time with monotonous regularity, which is more than can be said for all budding musicians attempting to follow them. Unfortunately, with conventional metronomes, it is all too easy for beginners to miss a beat at a fast tempo without even being aware of it. Meanwhile, the meticulous metronome ticks on relentlessly, generally unable to indicate visually that the luckless student has missed the boat, or more accurately, the beat, and strayed into the wrong part of the bar.

Most musicians will agree that metronome markings at the head of a piece of music serve only as a guide; but it is necessary, especially in group work, to learn to set or follow a strict tempo even if we eventually decide to go at our own sweet pace. Generally, most metronomes only sound and display a single beat regardless of time signature, but if we need a guide, why follow a one-legged one?

The unique Rhythm Setter displays *all* the basic beats of a bar, so there's no excuse for being 'down-beat' when you should be 'up-beat'.

2 Theme...

Some recent experiments with the versatile 4017 CMOS decade counter/divider IC, suggested this personal audio/visual metronome, which uses a separate LED indicator for each individual beat in most of the usual time signatures. In other words, it gives a bar's worth of notes at a selected time signature and speed, repeating them in sequence for as long as the musician desires (or his neighbours can stand!).

For audible indication, the loudspeaker gives a click per beat, with the first beat of each bar getting a rhythmical accent to distinguish it from the succeeding beats.

Fig. 10.1 Rhythm Setter front panel layout

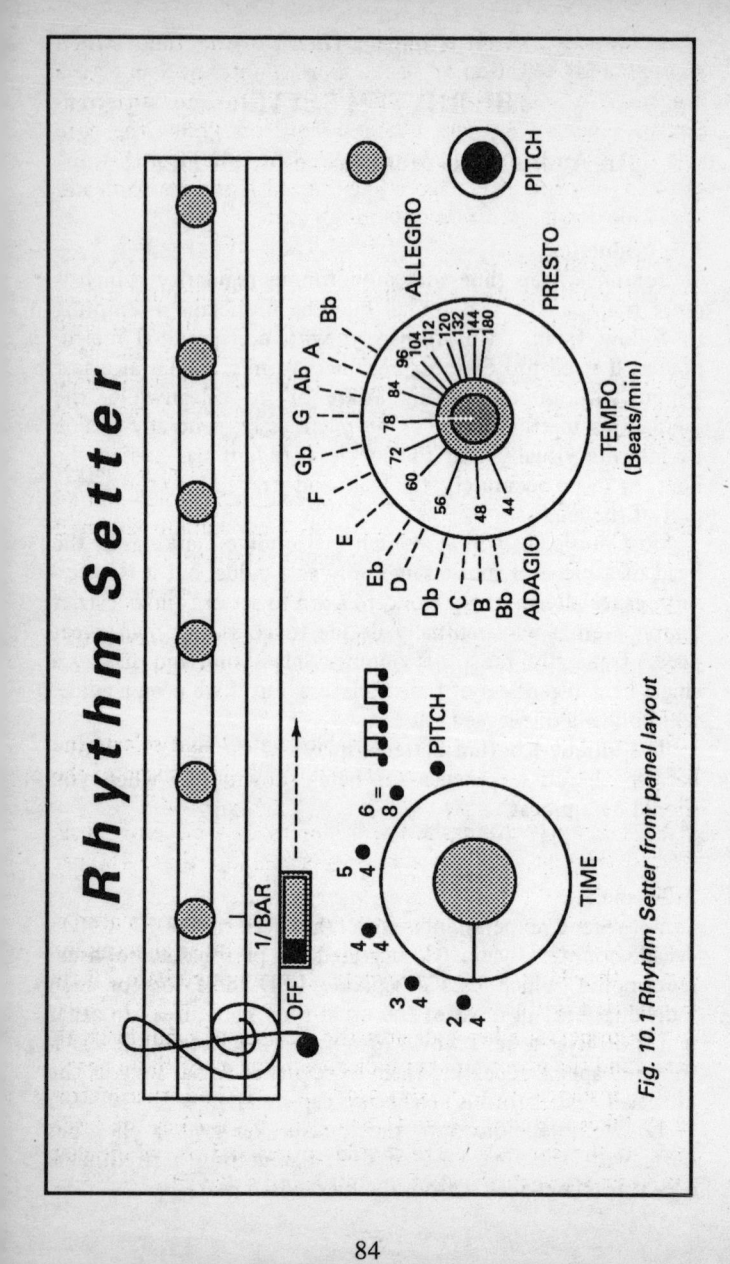

As an added extra, a 'pitch' position of the 'time' switch gives a handy selection of all the musical notes over an octave range. This is especially useful to pitch notes for unaccompanied singing. As most choral conductors know, the note one sounds on a piano isn't necessarily the same note that one gives to the choir after negotiating the width of a platform and knocking down a few music stands en route.

3 And Variations

The 4017 integrated circuit has ten decoded outputs, but as rhythms in nine-eight time are not too common it was decided to limit the display to a maximum of six beats; i.e. one bar's worth of quavers in six-eight time. However, if desired, three more LEDs and one more switch position could easily be added to include nine-eight compound rhythm.

A useful variation on this personal rhythm setter is a larger 'music workshop' model using an m.e.s. bulb display and an amplified beat. The basic circuit has been suitably adapted along these lines and used for class demonstrations.

4 First Movement — Modus operandi

The operating principle is straightforward, using only a trio of controls as shown on the front panel (Figure 10.1).

Tempo (beats/min)

This is simply a continuously-variable control that selects the ticking speed of the loudspeakers and the flashing speed of the LEDs.

Most tempo markings at the beginning of a piece of music indicate beat speeds from about 44 per minute up to 180 per minute (e.g. ♩ = 120).

If you are not too familiar with musical time signatures, the two numbers at the beginning of a musical work or section that look like a mathematical fraction, perhaps a little explanation is necessary.

The upper number indicates the number of main beats to a bar, and the lower number indicates the value of those beats.

For instance, in 2/4, 3/4, 4/4, and 5/4 time, this metronome takes a quarter note (1/4) or crochet (♩) as the basic beat, so the speed selected should correspond to the crochet tempo shown at the head of the music to be played.

However, in 6/8 time, we have two choices: we can either select the six quavers (1/8 note) in the bar or choose to indicate two slower dotted crochets, depending on the tempo (i.e. ♫♫ ♫♫ or ♩. ♩.).

Time

The Time control is a 6-way rotary switch that selects the number of beats in a bar, and hence the number of LEDs that will light in sequence.

The position of the switch must be selected to correspond with the time signature given before the first bar of the music:

2/4 gives two basic beats to a bar and the first two LEDs will light sequentially.
This position can be used for either two-four time (two crochet beats: ♩♩) or for fast six-eight time, to indicate two dotted crochets (♩. ♩.) in compound time.

3/4 gives three crochet beats (♩♩♩) to the bar and three LEDs light in sequence to display these. Alternatively, this position could also be used for three-eight time, three quavers in a bar (♫♫), providing the Tempo control is set to the tempo indicated for a quaver on the music to be played or sung.

4/4 indicates four crochets (♩♩♩♩) to the bar and four LEDs light. A breezy march tempo is 120 crochets to the minute (♩ = 120).

5/4 with five crochets to the bar (♩♩♩♩♩) is not often used, but is given here as the fifth LED is already included for the six-eight position.

6/8 is the odd position, being the only one that gives quavers (eighth notes) as the basic beat. In slow six-eight time, beating two basic dotted-crochets to the bar can look like an action replay; often so embarrassingly slow that the poor conductor doesn't know what to do with his hands!

This metronome position solves the problem by splitting each of the two dotted-crochets into its three quavers, enabling the conductor to fill in time by beating six to the bar (♫♩ ♫♩). In reality, the beat speeds up by a ratio of 3:1, and although we use the same Tempo scale in this position, it actually indicates the number of dotted crochets per minute. We must remember that the beats represent quavers in this compound time (six LEDs = six quavers).

Pitch The pitch facility uses one position of the Time switch and although it speeds up the beats/min into the audio range it is not strictly a time position. It permits the notes of a musical octave to be sounded quietly as a convenient means of pitching a note for say, unaccompanied singing. More volume could easily be obtained by adding a single transistor amplifier stage.

1/Bar

This is an all-purpose position, where only the first LED in the bar lights. It is usually suitable for fast tempi, where a conductor, either suffering from nervous exhaustion or in danger of being mistaken for a runaway windmill, gives up beating the individual rhythm notes and settles for one beat in a bar.

If it sounds confusing, this point is often clarified by the speed marking on the music copy, which instead of indicating, say (♩ = 180), in three-four time reads (♩. = 60). At this speed, most conductors happily break into one beat per second instead of three.

5 Second Movement – Opus 'Technicus' (Figure 10.2)

The popular decoded decade counter 4017 (IC2) is the heart of the instrument, used here as a divide by 'one to six' circuit depending on the position of the Time select switch S1. The basic operation can be seen from the block diagram, Figure 10.3.

The beat generator IC1 uses a 555 timer IC in free-run mode, the beat speed being governed by the Tempo control

Fig. 10.2 Rhythm Setter (circuit)

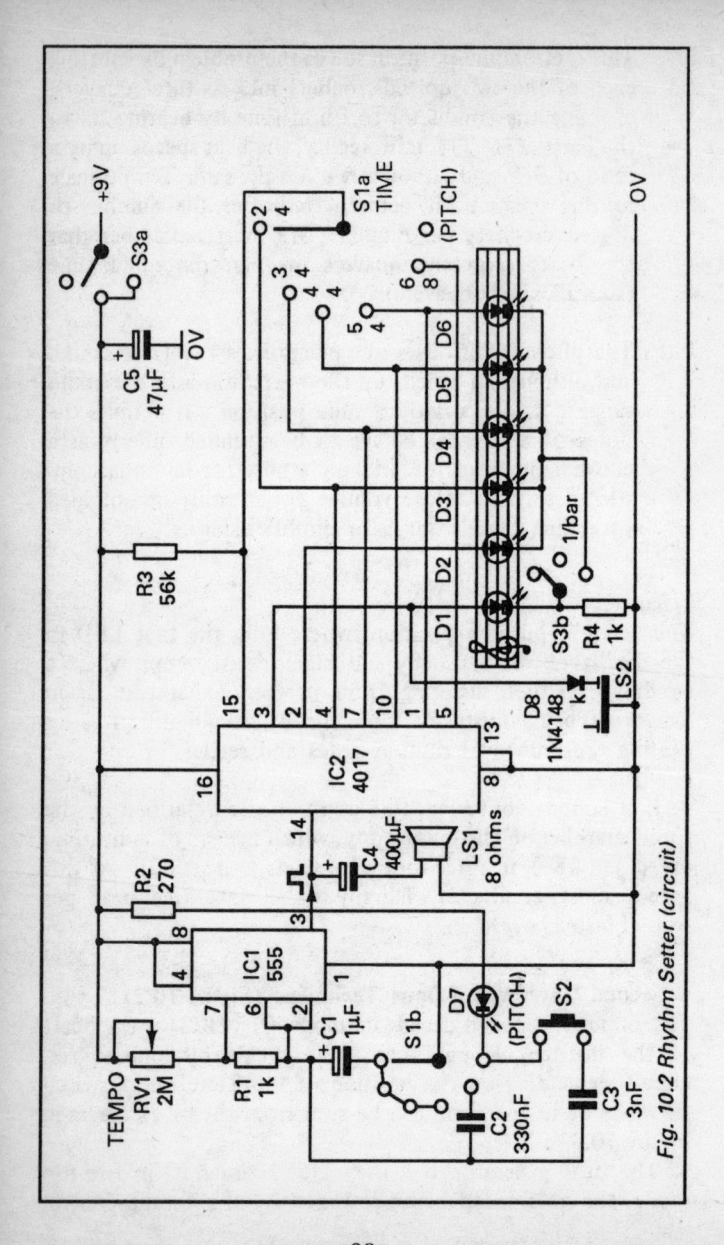

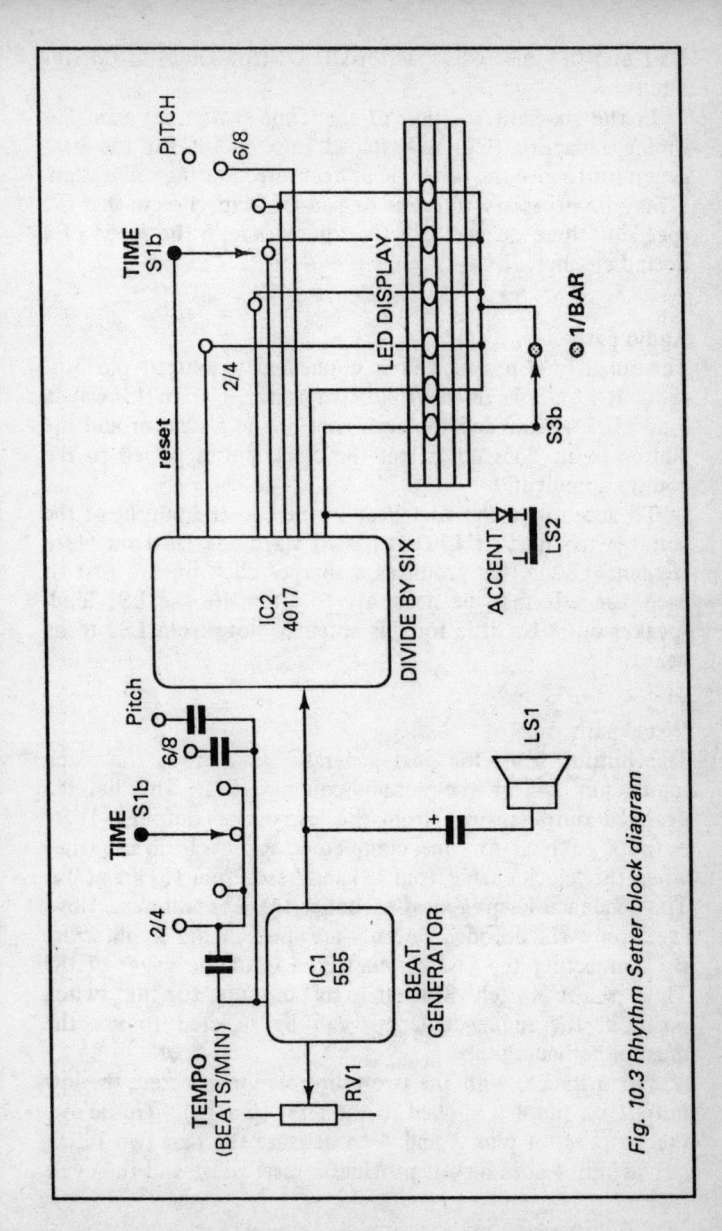

Fig. 10.3 Rhythm Setter block diagram

RV1 and the selected capacitor C1, C2 (for Time) or C3 (for Pitch).

In the six-eight position of the Time switch S1, a smaller timing capacitor (C2) is switched into circuit for the beat generator to give the faster (× 3) beat rate. During calibration, it may be necessary to select or pad this capacitor so that the speed of three quavers exactly coincides with the speed of a dotted crochet.

Audio path

The small loudspeaker LS1 is connected to output pin 3 of timer IC1 via a large electrolytic capacitor, C4. In this way, a loud click is sounded for each beat of the generator and the audio circuit does not affect the clock pulses passed to the counter circuit IC2.

To accentuate the first beat in the bar, the output of the counter to the first LED is passed via diode D8 to a piezo element LS2. This produces a sharper click on the first of each bar. It may be necessary to attenuate the LS1 loudspeaker output a little for this accented note from LS2 to be heard.

Visual path

The output from the beat generator also drives the clock input, pin 14, of the decade counter IC2. This has ten decoded outputs apart from the 'carry one' output. These outputs go high for one complete clock cycle in sequence when the 'clock enable' (pin 13) and 'reset' (pin 15) are at 0V. The sequence keeps repeating under these conditions. However, only six decoded outputs are used in this application. By connecting the reset input (pin 15) to the wiper of the Time select switch S1a, and six outputs to the switch position, the sequence length can be selected to suit the musical bar length.

For instance, with the two-four position selected, the low output on pin 4 is applied to the reset (pin 15). This allows the outputs on pins 3 and 4 to activate the first two LEDs, but as pin 4 goes high it provides a reset pulse and the cycle repeats.

Fig. 10.4 Rhythm Setter (layout)

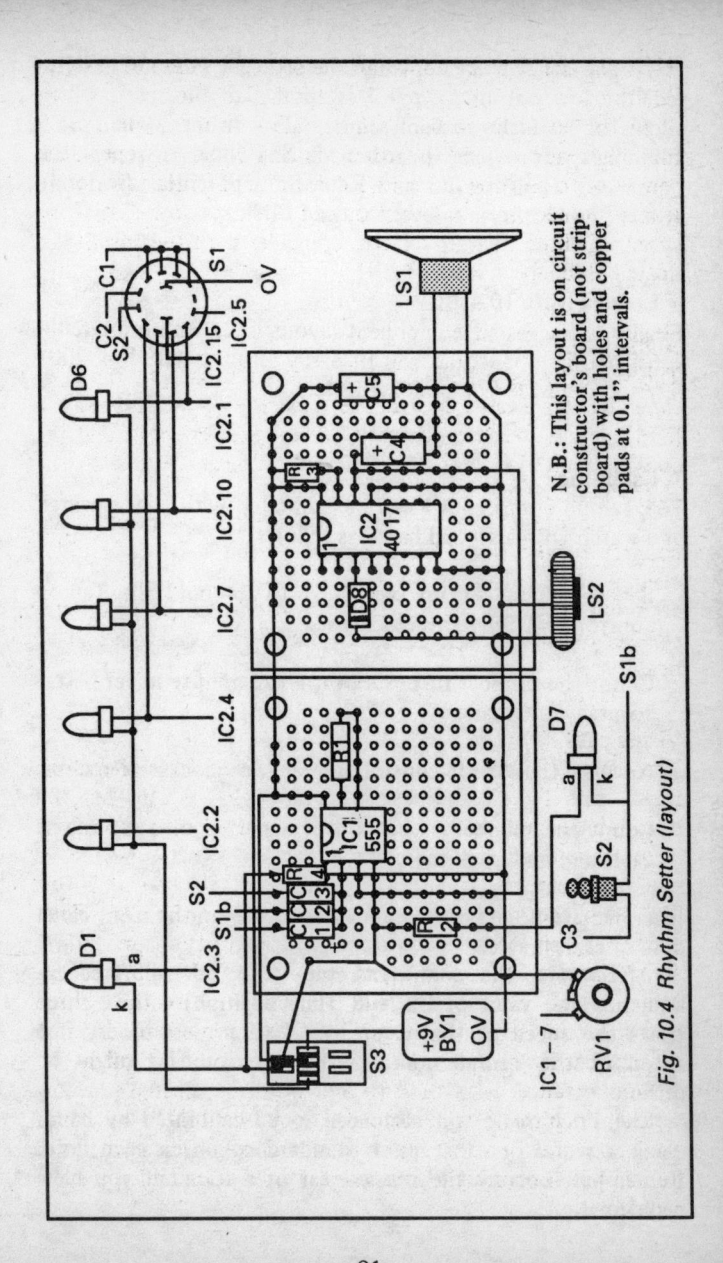

At the other extreme, when the six-eight position is selected, the low output on pin 5 is applied to the reset, which allows all six LEDs to light sequentially. In turn, when pin 5 goes high it provides the reset via S1a wiper to repeat the sequence. Resistors R3 and R4 form a potential divider to give the correct output level for the LEDs.

6 Coda (Figure 10.4)

Finally, a suggested component layout is shown on a circuit-constructor's board that will fit easily into the ABS Box 2005 referred to in the Components List.

7 Calibration

The Tempo control RV1 can be calibrated with a stop watch or a watch with a second hand, as follows:

Set the Time control switch S1 to 2/4 and the Tempo control RV1 fully anti-clockwise.

Count the number of beats over a full minute to get better accuracy and mark.

Advance the Tempo control progressively clockwise in steps.

Count the number of beats over a period of a minute at each step and mark these.

Intermediate points can be added as shown on the front panel dial, or as desirable.

Afterwards, the compound time can be calibrated by adjusting the value of C2 (6/8 Time position) to give three times the speed of that given by C1 (don't use a very fast speed of the Tempo control otherwise counting might be difficult).

The Pitch scale will also need to be calibrated by hand, using a piano or other pitch standard to check each note. Remember, borrow the musical ear of a friend if you have problems!

Components, Rhythm Setter

Resistors
R1   1k
R2   270 ohms
R3   56k
R4   1k

Potentiometers
RV1   2M (linear)

Capacitors
C1   1µF 10V
C2   330nF plastic foil
C3   3nF plastic foil
C4   400µF 10V
C5   47µF 10V

Semiconductors
D1 to D7   TIL20 LEDs
D8   1N4148 silicon diode
IC1   NE555CP timer IC
IC2   CD4017 CMOS decade counter/divider

Loudspeakers
LS1   8 ohms
LS2   piezo element

Switches
S1   double-pole 6-way rotary
S2   pushbutton (non-locking)
S3   double-pole 3-way miniature slider

Battery
BY1   9V PP3

Miscellaneous
Suitable case (ABS Box 2005 150 × 80 × 50mm approx.), battery connector, constructor's printed-circuit board, connecting wire, control knobs, etc.

Chapter 11

THE GLIDAPHONE

Stringed instruments and trombones have one thing in common — glissando! This is the ability to glide from one note to another with a smooth variation of pitch. Apart from Hawaiian guitars, which seem to thrive on it, glissando is used very sparingly by composers and only for special effects: to imitate the roar of a lion, a dramatic effect, a jazz idiom, or a musical joke — a trombone glissando was almost a stock effect in old comedy films when some character slipped on a banana skin. Fortunately, an instrument that is capable of glissando brings with it the ability to produce a most natural and pleasing vibrato.

These musical effects are exploited in the Glidaphone circuit. Over the playing range, the pitch of the notes is controlled by a tuning arm that can be rotated through 180 degrees to span an octave and a half or more — sufficient for most melodies. The notes are sounded by a fingertip touch action on a metal plate labelled with the notes of the musical scale. For greater range, the basic range of notes can be switched up or down by an octave as desired.

1 Mechanical Construction (Figure 11.1)

A suggested front panel layout of the Glidaphone is shown in Figure 11.1. The triangular shaped panel allows the controls to be conveniently situated for the player when handheld. With the left-hand supporting the bottom corner, the fingers are available to vary the intensity and octave pitch controls. The right hand is then free to rotate the tuning control arm. Left-handers can construct a mirror-image version. The prototype was made mainly from 3-ply. One diagonal cut across a 15-in square piece of plywood or hardboard can supply both the front and the base panel. For the sides of the base, shown in Figure 11.2, you will need a 4 ft length of 2-in × ½-in batten. Cut four lengths as shown, and tack and glue these to the edge of the base. When the glue is set, round-off the corners with a flat file.

Fig. 11.1 Glidaphone front panel

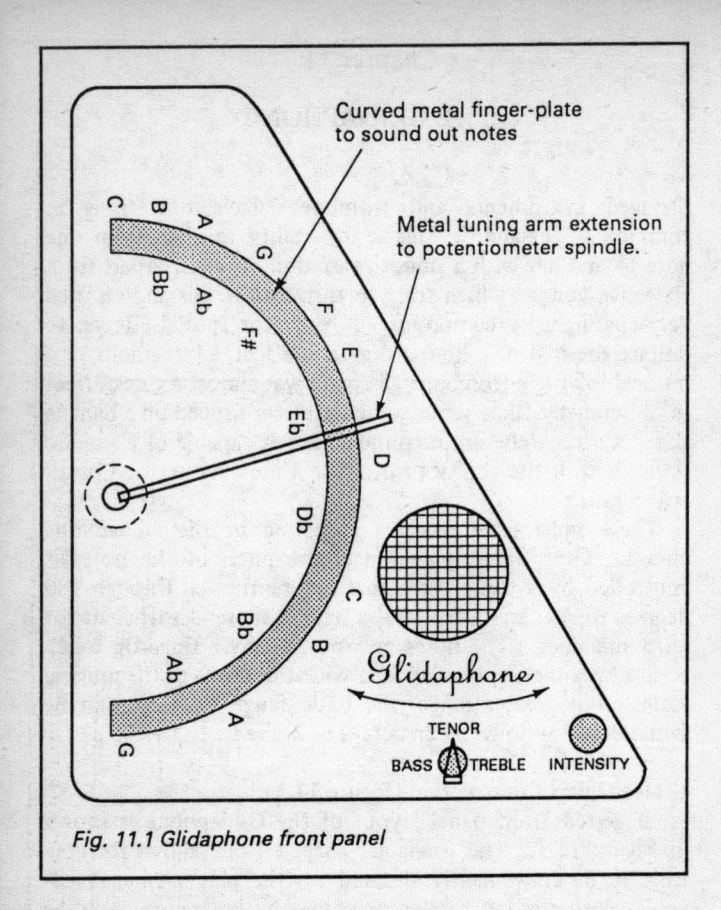

Drill four small holes in the edges of the front panel and fit it to the base using woodscrews. Round-off the corners to match the base taking care not to split the wood. Now that the case has taken shape, remove the cover and cut the holes for the three control spindles and the loudspeaker. As shown, the curved-metal finger-plate is semi-circular. This can be made from aluminium, stainless-steel plate or even baking foil glued to the panel. An electrical connection is needed through to the circuit. This can be a small bolt, solder-tag and nut through one end of the metal

Fig. 11.2 Glidaphone (construction)

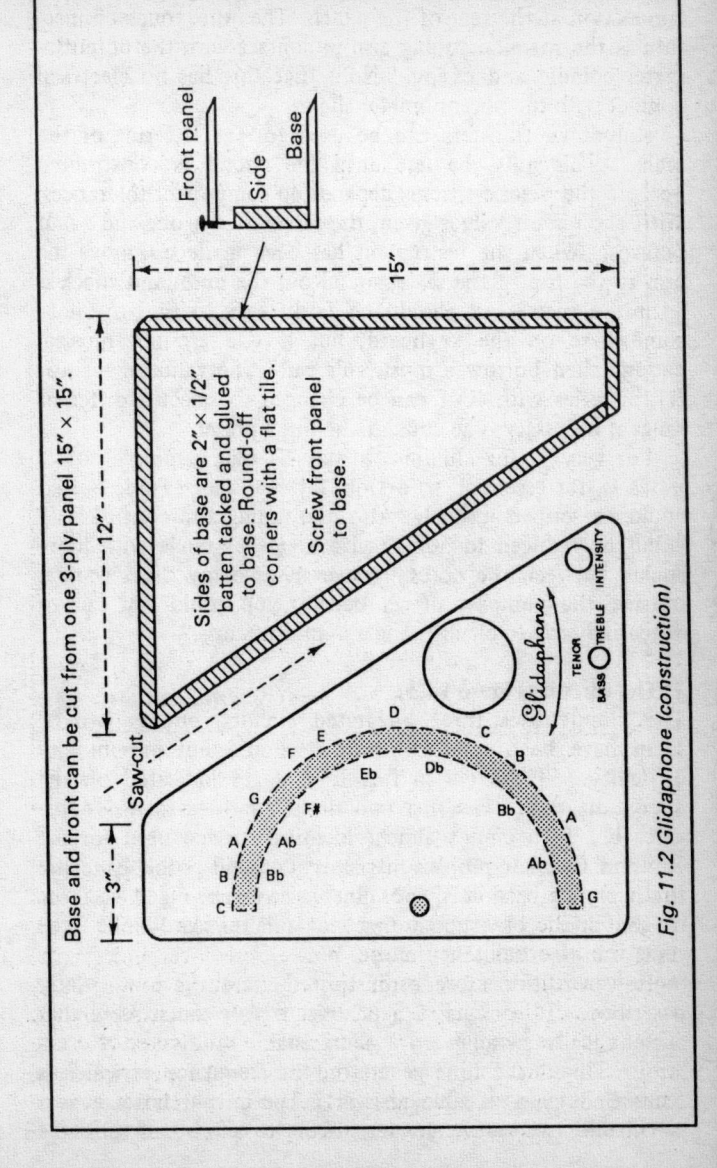

strip, or a brass woodscrew of sufficient length to solder a connection at the rear of the panel. The other touch connection is the metallic tuning arm in contact with the potentiometer spindle and casing. Note that this has no electrical contact with the potentiometer slider.

Rub-down transfers can be used for the lettering of the scale. This must be left until the circuit is constructed because the precise spacing depends on component tolerances. With the circuit values given, the range is over one-and-a-half octaves. When the instrument has been made up, move the arm to the top of the scale, sound out the note, and check it against a piano. It should be fairly easy to find the note somewhere on the keyboard, but if you are not musical enough then borrow a musician's ear. The value of resistor R1 in series with RV1 can be changed to select a different range if necessary; you could make this a preset.

For very young children, it may be that naming absolute notes is not essential, so a Doh, ray, me scale (Tonic Sol-fa) could be scribed instead. Also, the tuning arm control RV1 could be reduced to 50k to give a smaller range with more spread between the notes. However, as many tunes wander outside the compass of an octave, you could run out of notes and have to change ranges in midstream!

2 The Circuit (Figure 11.3)

The circuit uses three integrated circuits, but as two of them have been used before this should not present any difficulty. There are in fact four stages indicated on the circuit diagram. The first two share the 4011 quad NAND-gate, IC1 in a circuit almost identical to that used for the Rhythm Checker project. By now, you will probably realise that a simple version of the Glidaphone can be made from the rhythm checker by substituting a variable resistor for the fixed resistor in the oscillator circuit.

However, this more sophisticated circuit is now briefly described. It consists of a Schmitt trigger touch stage that automatically switches on a gated astable multivibrator oscillator. This is the tone generator, the frequency of which is controlled by a variable resistor. The output note feeds a switchable two-octave divider circuit to pitch the range of

Fig. 11.3 Glidaphone (circuit)

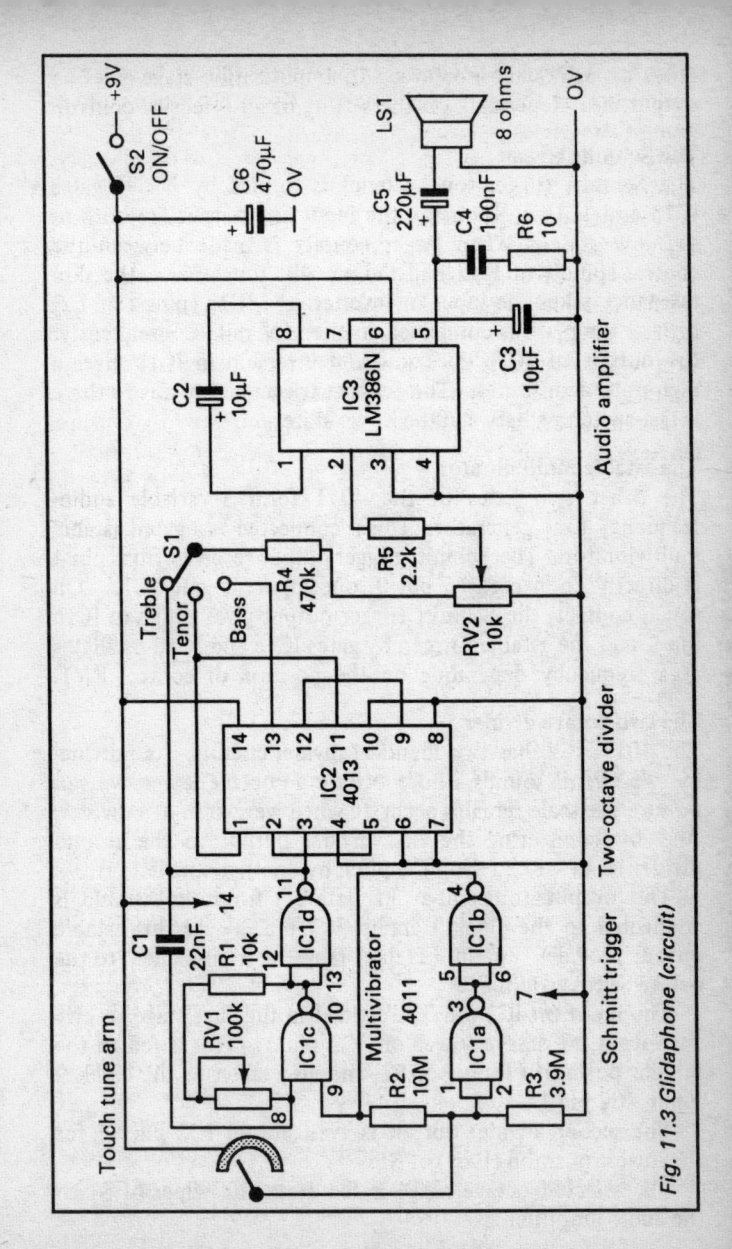

notes up or down an octave. An amplification stage gives an output level depending on the setting of an intensity control.

The Schmitt trigger

The Schmitt trigger touch circuit is formed by NAND-gates IC1a and IC1b. Normally, the input and output are both in the low state. When finger contact is made between the control spindle of RV1 and the metallic touch-plate, the skin resistance takes the input of inverter gate IC1a (pins 1 and 2) high as the plate is connected to the +9V rail. Consequently, the output of IC1a goes low and inversion in IC1b gives a high output on pin 4. The Schmitt trigger action means there is fast switching between these two states.

The astable multivibrator

The other two gates of the 4011 form a variable audio-frequency tone generator. This is connected as a gated astable multivibrator. The Schmitt trigger output received from pin 4 is directly connected to pin 9, one input of gate IC1c. On touch contact, the Schmitt trigger output gives a high to IC1c pin 9 and the astable formed by gates IC1c and IC1d oscillates at a frequency depending on the position of control RV1.

The two-octave divider

The 4013, IC2, has two identical divider circuits. As a divide-by-two circuit sounds a note exactly an octave lower, we can be sure the scale remains accurate when we switch in a divider. And by connecting the first divider output to the second divider input we can drop the pitch by another octave.

The output (one-and-a-half octaves) from the astable is connected to the clock 1 input, IC2 pin 3. As this astable output provides our top (Treble) range, it is also taken to the octave select switch S1.

The input on IC2 pin 3 is divided in the first stage to give an output an octave lower on IC2 pin 1. This is fed to the middle position (Tenor) of S1, and also taken to the clock 2 input, IC2 pin 11.

The second divider output is available on IC2 pin 13 for the lowest position (Bass) of S1.

The selected octave range is fed from the wiper of S1 to the audio amplifier.

Audio amplifier

The series resistor R4 limits the output to prevent overloading the amplifier stage. This stage is identical to that used in the Elexylophone project except for component numbering. The input to the LM386N amplifier is controlled by the Intensity control RV2. If this stage has not already been made as a separate module, then refer to Figure 9.4 for constructional details of the amplifier. The output is about 250mW.

Fine! That's not a compliment. It's the Italian word to say the musical piece is finished. So! What do we do with it now?

3 Shake it, slide it!

Sound out a note by holding the tuning-arm with your right hand and touching your forefinger on the metal plate. Musically, it sounds a bit bare. Sound it again, but this time wave your left hand across the speaker at about five times per second. The slow shake to the note makes it more pleasant. This stretching of the sound waves (Doppler effect) alters the pitch or frequency of the notes. We know that musicians call this vibrato. Try to get the same effect by a slight up and down movement of the tuning arm a few times per second. It takes a little practice to get it sounding musical and more practice to add it to a tune. It's better to add vibrato to the longer notes of a melody. Watch famous big band trombonists like Don Lusher. They start the note straight then add a little judicious movement of the slide to warm the sound as it progresses.

Try some gliding effects for instance from C down to G and back again. Use them sparingly! Trombonists are about the only musicians who can succeed by letting things slide!

Components, Glidaphone

Resistors

R1	10k (fixed or preset)
R2	10M
R3	3.9M
R4	470k
R5	2.2k
R6	10 ohms

Potentiometers
RV1   100k
RV2   10k with ON/OFF switch S2

Capacitors
C1   22nF plastic foil
C2   10µF 10V
C3   10µF 10V
C4   100nF plastic foil
C5   220µF 10V
C6   470µF 10V

Semiconductors
IC1   4011 Quad 2-input NAND-gate
IC2   4013 divider
IC3   LM386N audio amplifier

Loudspeaker
LS1   8 ohms

Switches
S1   Single-pole, 3-way
S2   S.P.S.T. on RV2

Miscellaneous
Box, metal finger touch-plate, metal extension arm, stripboard, 9V battery, wire, etc.

Chapter 12

THE CHORDMAKER

Most of the project chapters have dealt with a musical characteristic; for example, bar-lengths, note values, rhythms, tempo, pitch, intensity and quality. The instrument projects have concentrated on simple single-tone or monophonic sounds, but multi-tone or polyphonic alternatives have been mentioned. Although the study of harmony — two or more sounds heard at the same time — is too advanced for this book, the construction of an instrument that will produce the primary chords of a major scale provides a good starter for students.

Chords are formed by two or more notes built up at intervals of a third; i.e. by skipping a letter name; for example, C to E, E to G. The lowest note that the thirds are built on is called the *root*. However, as the letter names A to G repeat after an octave, the position of the notes in a chord are often changed around, for instance, to get a smoother movement of parts betwen chords. If the root is kept as the lowest note (the bass note) the chord is said to be in root position.

Common chords can be built on all degrees of a major scale by adding two notes at intervals of a third (a chord with three notes is often referred to as a *triad*). However, only those chords built on the first note of the scale, the fourth note and the fifth note are major chords. That is, the interval of the third from the root (e.g. C to E) has four semitones. You can verify this on a piano or with the Electronic Sol-fa layout (Figure 4.1). The diagram also gives the names of the root notes of these major chords. The chord with the first or keynote of the scale as root is called the *tonic*.

The chord on the fourth degree of the scale is the *subdominant* and the chord on the fifth degree is called the *dominant*.

With these three *primary* chords, all the notes of the diatonic scale can be simply harmonised. So, as long as your melody does not stray out of the key of C major it can be harmonised by the Chordmaker. A tune often starts on one

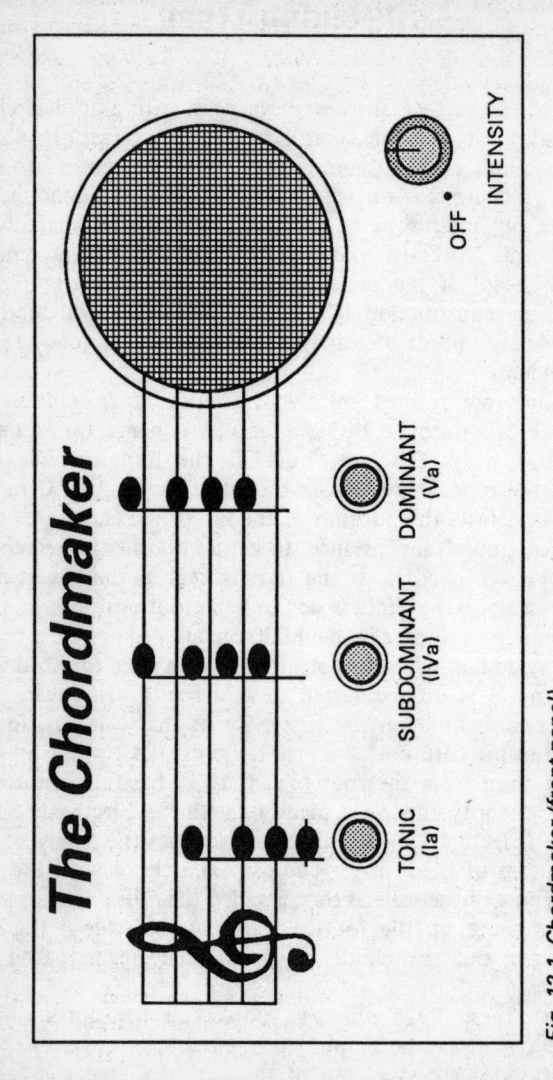

Fig. 12.1 Chordmaker (front panel)

of the notes of the tonic chord, but you can be almost certain it will finish on the tonic note. So if you want to use the Chordmaker (which is in C major) to harmonise a song, it must be sung or played in the key of C. Easy enough with an instrument, but more difficult to pitch with the voice! To be sure you pitch your song in C, here's how to do it. First play the tonic chord. Then imagine the C in the chord to be the last note of your melody.

Other keys could be chosen, and more interesting chords added if desired, but the three major chords in the key of C will prove the point. If you look at the front panel (Figure 12.1) of the Chordmaker you will see that all the notes of the C major scale are present somewhere in the three chords. In four-part harmony, it is necessary to double one of the notes of a chord. In root position, the best note to double is the bass. That's the reason for the extra note at the top of each of the three chords. In three-part harmony doubling is generally omitted, otherwise the parts are restricted to two notes.

1 The Circuit (Figure 12.2)

If you have made up the handbells or the Elexylophone, you will have no problems in understanding the circuit of the Chordmaker. For the tone generators I decided for the tried and trusty Hartley circuit. To form the three chords, ten are needed. This may sound extravagent as not more than four notes are needed at any one time. This could be done with four oscillators, by picking a note off each one, but the Hartleys are very stable, have interesting possibilities, and are easy to construct. Another bonus is that they can be made up on the wider track stripboard — much easier on the eyes than the 0.1in version needed for CMOS ICs.

The oscillator circuits are almost identical, but the way they are selected for the chords needs a little explanation. The output amplifier for the prototype was again the LM386N used for other projects, but any amplifier can be used providing it does not introduce intermodulation distortion. With the single note projects, a little distortion is not so noticeable, but with two or more notes together, mixing occurs with some peculiar results. The oscillators produce

Fig. 12.2 Chordmaker (circuit)

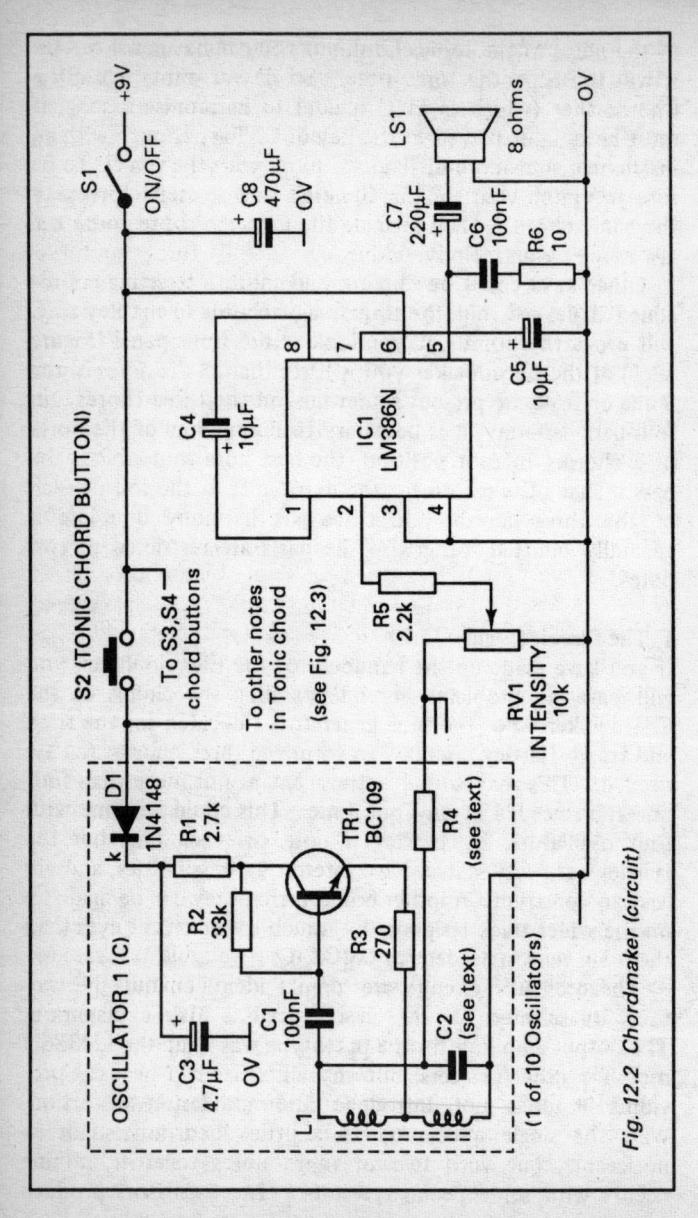

some very healthy signals so the outputs may need to be attenuated to match each other and to prevent overloading the amplifier.

Oscillators

The Hartley circuit of TR1 is identical to that used in previous projects. The tapped coil L1 does not need to be wound on a Ferroxcube core to get frequency stability throughout the note length, as each circuit is supplied with a constant collector voltage from the +9V rail for oscillation; i.e. the decay time of C3 is very short when the chord button is released. The frequency of the note will depend on L1 and C2. If air-cored coils are used, more turns will be needed to get the required inductance as explained in the Elexylophone project. A metal or ferrite slug can be screwed into the coil for fine tuning.

The resistors R4 on each oscillator can be finally selected (or adjusted) to match individual outputs, so make sure they are accessible. These can vary over values of several 100k to get even outputs from each note, depending on components.

One side of the three chord pushbuttons S2, S3 and S4 is connected to the +9V rail. The other side of each pushbutton connects to the four oscillators in its particular chord. In the circuit diagram of Figure 12.2 only oscillator 1 is shown in detail. This is the C or tonic oscillator, and is energised by the S2 pushbutton which provides collector voltage for TR1 via diode D1. Similarly, the S2 pushbutton is also connected via diodes to the other notes in the tonic chord (E, G, C^1).

2 Construction and Interconnections

There is nothing unusual about the box for the Chordmaker. It can be made up from plywood, or a suitable plastic container can be pressed into service. Its size will mainly depend on the space needed for the ten oscillators. Maplin produce a range of plastic project boxes of different shapes and sizes. The ABS Box MB6 with internal dimensions 214 × 143 × 61mm should be large enough for the basic instrument if you do not have a suitable box to hand. The loudspeaker can be mounted behind the front panel (Figure 12.1), and the cut-out will depend on the size. Four holes need to be drilled to

Fig. 12.3 Chordmaker (connections)

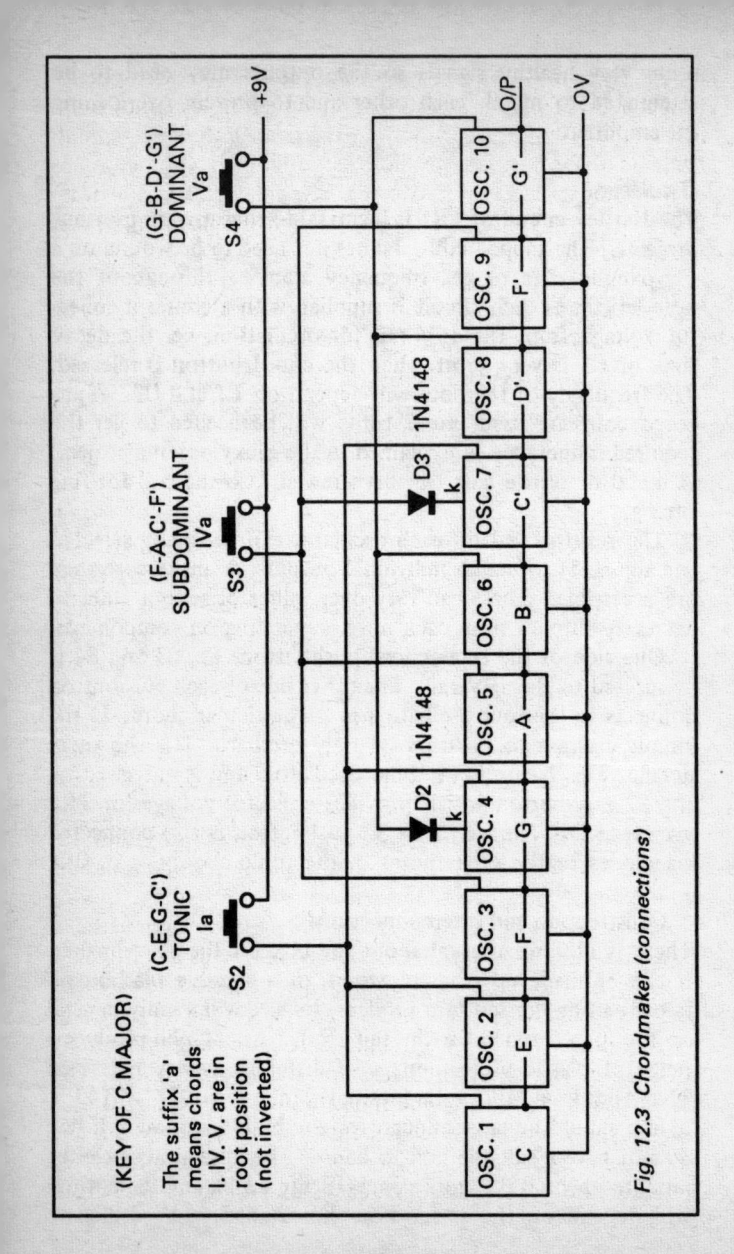

mount the three chord pushbuttons and the Intensity potentiometer RV1. The notes shown on the treble staff not only make the front panel more attractive, but also remind students of the notes as the chords are sounded.

All interconnections between the oscillators are shown in Figure 12.3. There are three standard connections to all oscillators:

a connection to D1 from the appropriate chord pushbutton to energise the oscillator;

a common output connection from R4 of all oscillators to the audio amplifier input control RV1;

a common earth connection to the 0V rail.

In addition, there are two more supply diodes fitted because two of the notes, G and C^1, are common to two chords: G appears in both the tonic and dominant chords, while C^1 appears in the tonic and subdominant chords. Diode D2 is connected to the cathode end (k) of D1 on oscillator 4, and D3 is connected to the cathode end (k) of D1 on oscillator 7. Digressing slightly, these link notes give us a choice of chords when harmonising these notes of the scale. You will easily discover which chords are the most suitable. Note that the tonic chord to harmonise C sounds much more final than the subdominant chord.

Some of the diodes may seem to be unnecessary, but are needed if the Chordmaker is extended. They provide isolation when chords share notes, and allow a selected group of oscillators to be energised from a common pushbutton without giving up their independence. For instance, this circuit could be used as the basis of a polyphonic keyboard instrument with an oscillator for each note. Keying on the cathode side of the diode D1 would sound each note singly, without bringing in the other oscillators of the chord group. The chord pushbuttons could be mounted on the keyboard as a simple chord organ to accompany a melody played on the keyboard. Alternatively, for the accomplished player, all notes of the keyboard are available to produce a full organ effect.

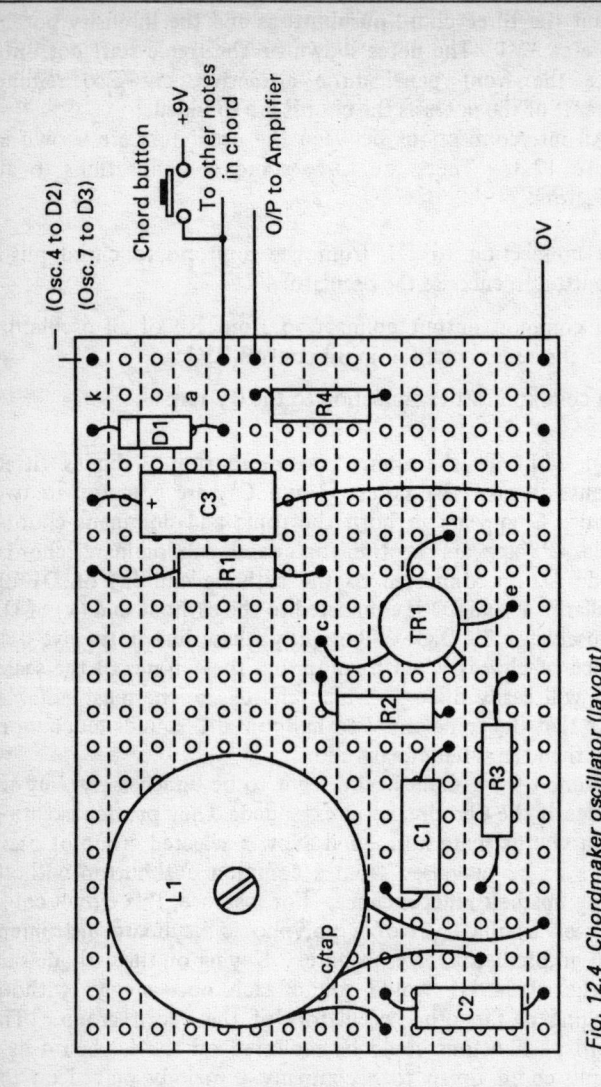

Fig. 12.4 Chordmaker oscillator (layout)

The layout for one oscillator is shown in Figure 12.4. The layout is identical for all others except that the extra diodes D2 (for oscillator 4) and D3 (for oscillator 7) are connected (cathode end) to the top right-hand corner hole of the board.

Components, Chordmaker

Per Oscillator

Resistors

R1	2.7k
R2	33k
R3	270 ohms
R4	100k ... 470k (see text)

Capacitors

C1	100nF plastic foil
C2	47nF to 500nF (see text)
C3	4.7µF 10V

Coils

L1	400–0–400 turns 39swg on Ferroxcube core or 1500–0–1500 turns, air-cored.

Semiconductors

TR1	BC109
D1	1N4148
D2	1N4148 (Oscillator 4 only)
D3	1N4148 (Oscillator 7 only)

Amplifier

Resistors

R5	2.2k
R6	10 ohms

Potentiometers

RV1	10k (with S1) Intensity and ON/OFF

Capacitors
C4	10µF 10V
C5	10µF 10V
C6	100nF plastic foil
C7	220µF 10V
C8	470µF 10V

Semiconductors
IC1	LM386N

Loudspeaker
LS1	8 ohms

Switches
S1	S.P.S.T. on RV1
S2	S.P.S.T. push to make
S3	S.P.S.T. push to make
S4	S.P.S.T. push to make

Miscellaneous
Suitable case, stripboard, 9V battery, insulated wire, 39swg enamelled wire for winding coils.

List of Figures

No.	Title	Page
2.1	Electronic circuit component chart	8-10
2.2	The Great Staff	15
2.3	Notes and bars	16
3.1	Rhythm Checker bar frame	20
3.2	Time signature, notes and rests	22
3.3	Rhythm Checker indicator circuit Mk.I	23
3.4	Rhythm Checker indicator circuit Mk.II	25
3.5	Rhythm Checker touchpad	28
3.6	Finger-tapping Rhythmboard	29
4.1	Electronic Sol-fa (layout)	32
4.2	Electronic Sol-fa (circuit)	34
4.3a	Component side of stripboard	36
4.3b	Underside of stripboard	37
5.1	Dynamics Ranger (front panel)	42
5.2	Dynamics Ranger (circuit)	43
6.1	Tune-up Box (layout)	50
6.2	Tune-up Box (circuit)	52
7.1	Melody Ranger (keyboard layout)	58
7.2	Melody Ranger (circuit)	60
7.3	Melody Ranger (key mounting details)	61
7.4	Vibrato oscillator	64
8.1	Handbell construction	68
8.2	Handbell (circuit)	70
8.3	Handbell (layout)	72
9.1	Elexylophone (keyboard layout)	74
9.2	Elexylophone (circuit)	75
9.3	Elexylophone (oscillator layout)	78
9.4	Elexylophone (amplifier layout)	79
10.1	Rhythm Setter (front panel)	84
10.2	Rhythm Setter (circuit)	88
10.3	Rhythm Setter (block diagram)	89
10.4	Rhythm Setter (layout)	91
11.1	Glidaphone (front panel)	96
11.2	Glidaphone (construction)	97
11.3	Glidaphone (circuit)	99

No.	Title	Page
12.1	Chordmaker (front panel)	104
12.2	Chordmaker (circuit)	106
12.3	Chordmaker (connections)	108
12.4	Chordmaker oscillator (layout)	110

Notes

Notes

Notes

Please note following is a list of other titles that are available in our range of Radio, Electronics and Computer books.

These should be available from all good Booksellers, Radio Component Dealers and Mail Order Companies.

However, should you experience difficulty in obtaining any title in your area, then please write directly to the Publisher enclosing payment to cover the cost of the book plus adequate postage.

If you would like a complete catalogue of our entire range of Radio, Electronics and Computer Books then please send a Stamped Addressed Envelope to:

BERNARD BABANI (publishing) LTD
THE GRAMPIANS
SHEPHERDS BUSH ROAD
LONDON W6 7NF
ENGLAND

160	Coil Design and Construction Manual	£2.50
227	Beginners Guide to Building Electronic Projects	£1.95
BP28	Resistor Selection Handbook	£0.60
BP36	50 Circuits Using Germanium Silicon & Zener Diodes	£1.95
BP37	50 Projects Using Relays, SCRs and TRIACs	£2.95
BP39	50 (FET) Field Effect Transistor Projects	£2.95
BP42	50 Simple LED Circuits	£1.95
BP44	IC 555 Projects	£2.95
BP45	Projects in Opto-Electronics	£2.50
BP48	Electronic Projects for Beginners	£1.95
BP49	Popular Electronic Projects	£2.50
BP53	Practical Electronics Calculations & Formulae	£3.95
BP56	Electronic Security Devices	£2.95
BP58	50 Circuits Using 7400 Series IC's	£2.50
BP74	Electronic Music Projects	£2.95
BP76	Power Supply Projects	£2.50
BP78	Practical Computer Experiments	£1.75
BP80	Popular Electronic Circuits – Book 1	£2.95
BP84	Digital IC Projects	£1.95
BP85	International Transistor Equivalents Guide	£3.95
BP87	50 Simple LED Circuits – Book 2	£1.95
BP88	How to Use Op-amps	£2.95
BP90	Audio Projects	£2.50
BP92	Electronics Simplified – Crystal Set Construction	£1.75
BP94	Electronic Projects for Cars and Boats	£1.95
BP95	Model Railway Projects	£2.95
BP97	IC Projects for Beginners	£1.95
BP98	Popular Electronic Circuits – Book 2	£2.95
BP99	Mini-matrix Board Projects	£2.50
BP104	Electronic Science Projects	£2.95
BP105	Aerial Projects	£2.50
BP107	30 Solderless Breadboard Projects – Book 1	£2.95
BP110	How to Get Your Electronic Projects Working	£2.95
BP111	Audio	£3.95
BP112	A Z-80 Workshop Manual	£3.95
BP113	30 Solderless Breadboard Projects – Book 2	£2.25
BP115	The Pre-computer Book	£1.95
BP118	Practical Electronic Building Blocks – Book 2	£1.95
BP121	How to Design and Make Your Own PCB's	£2.50
BP122	Audio Amplifier Construction	£2.95
BP125	25 Simple amateur band Aerials	£1.95
BP126	BASIC & PASCAL in Parallel	£1.50
BP130	Micro Interfacing Circuits – Book 1	£2.75
BP131	Micro Interfacing Circuits – Book 2	£2.75
BP132	25 Simple SW Broadcast Band Aerials	£1.95
BP136	25 Simple Indoor and Window Aerials	£1.75
BP137	BASIC & FORTRAN in Parallel	£1.95
BP138	BASIC & FORTH in Parallel	£1.95
BP144	Further Practical Electronics Calculations & Formulae	£4.95
BP145	25 Simple Tropical and MW Band Aerials	£1.75
BP146	The Pre-BASIC Book	£2.95
BP147	An Introduction to 6502 Machine Code	£2.95
BP148	Computer Terminology Explained	£1.95
BP152	An Introduction to Z80 Machine Code	£2.95
BP171	Easy Add-on Projects for Amstrad CPC 464, 664, 6128 & MSX Computers	£2.95
BP174	More Advanced Electronic Music Projects	£2.95
BP176	A TV-DXers Handbook (Revised Edition)	£5.95
BP177	An Introduction to Computer Communications	£2.95
BP179	Electronic Circuits for the Computer Control of Robots	£2.95
BP180	Electronic Circuits for the Computer Control of Model Railways	£2.95
BP182	MIDI Projects	£2.95
BP184	An Introduction to 68000 Assembly Language	£2.95
BP187	A Practical Reference Guide to Word Processing on the Amstrad PCW8256 & PCW8512	£5.95
BP190	More Advanced Electronic Security Projects	£2.95
BP192	More Advanced Power Supply Projects	£2.95
BP193	LOGO for Beginners	£2.95
BP194	Modern Opto Device Projects	£2.95
BP195	An Introduction to Satellite Television	£5.95
BP196	BASIC & LOGO in Parallel	£2.95
BP197	An Introduction to the Amstrad PC's	£5.95
BP198	An Introduction to Antenna Theory	£2.95
BP230	A Concise Introduction to GEM	£2.95
BP232	A Concise Introduction to MS-DOS	£2.95
BP233	Electronic Hobbyists Handbook	£4.95
BP239	Getting the Most From Your Multimeter	£2.95
BP240	Remote Control Handbook	£3.95
BP243	BBC BASIC86 on the Amstrad PC's & IBM Compatibles – Book 1: Language	£3.95
BP244	BBC BASIC86 on the Amstrad PC's & IBM Compatibles – Book 2: Graphics and Disk Files	£3.95
BP245	Digital Audio Projects	£2.95
BP246	Musical Applications of the Atari ST's	£5.95
BP247	More Advanced MIDI Projects	£2.95
BP248	Test Equipment Construction	£2.95
BP249	More Advanced Test Equipment Construction	£3.50

Code	Title	Price
BP250	Programming in FORTRAN 77	£4.95
BP251	Computer Hobbyists Handbook	£5.95
BP254	From Atoms to Amperes	£3.50
BP255	International Radio Stations Guide (Revised 1991/92 Edition)	£5.95
BP256	An Introduction to Loudspeakers & Enclosure Design	£2.95
BP257	An Introduction to Amateur Radio	£3.50
BP258	Learning to Program in C	£4.95
BP259	A Concise Introduction to UNIX	£2.95
BP260	A Concise Introduction to OS/2	£2.95
BP261	A Concise Introduction to Lotus 1-2-3 (Revised Edition)	£3.95
BP262	A Concise Introduction to Wordperfect (Revised Edition)	£3.95
BP263	A Concise Introduction to dBASE	£3.95
BP264	A Concise Advanced User's Guide to MS-DOS (Revised Edition)	£3.95
BP265	More Advanced Uses of the Multimeter	£2.95
BP266	Electronic Modules and Systems for Beginners	£3.95
BP267	How to Use Oscilloscopes & Other Test Equipment	£3.50
BP269	An Introduction to Desktop Publishing	£5.95
BP270	A Concise Introduction to Symphony	£3.95
BP271	How to Expand, Modernise & Repair PC's & Compatibles	£4.95
BP272	Interfacing PC's and Compatibles	£3.95
BP273	Practical Electronic Sensors	£4.95
BP274	A Concise Introduction to SuperCalc5	£3.95
BP275	Simple Short Wave Receiver Construction	£3.95
BP276	Short Wave Superhet Receiver Construction	£2.95
BP277	High Power Audio Amplifier Construction	£3.95
BP278	Experimental Antenna Topics	£3.50
BP279	A Concise Introduction to Excel	£3.95
BP280	Getting the Most From Your PC's Hard Disk	£3.95
BP281	An Introduction to VHF/UHF for Radio Amateurs	£3.50
BP282	Understanding PC specifications	£3.95
BP283	A Concise Introduction to SmartWare II	£4.95
BP284	Programming in QuickBASIC	£4.95
BP285	A Beginners Guide to Modern Electronic Components	£3.95
BP286	A Reference Guide to Basic Electronics Terms	£5.95
BP287	A Reference Guide to Practical Electronics Terms	£5.95
BP288	A Concise Introduction to Windows3.0	£3.95
BP290	An Introduction to Amateur Communications Satellite	£3.95
BP291	A Concise Introduction to Ventura	£3.95
BP292	Public Address Loudspeaker Systems	£3.95
BP293	An Introduction to Radio Wave Propagation	£3.95
BP294	A Concise Introduction to Microsoft Works	£4.95
BP295	A Concise Introduction to Word for Windows	£4.95
BP297	Loudspeakers for Musicians	£3.95
BP298	A Concise Introduction to the Mac System & Finder	£3.95
BP299	Practical Electronic Filters	£4.95
BP300	Setting Up An Amateur Radio Station	£3.95
BP301	Antennas for VHF and UHF	£3.95
BP302	A Concise Users Guide to Lotus 1-2-3 Release 3.1	£3.95
BP303	Understanding PC Software	£4.95
BP304	Projects for Radio Amateurs and SWLs	£3.95
BP305	Learning CAD with AutoSketch	£4.95
BP307	A Concise Introduction to QuarkXPress	£4.95
BP308	A Concise Introduction to Word 4 on the Macintosh	£4.95
BP309	Preamplifier and Filter Circuits	£3.95
BP310	Acoustic Feedback – How to Avoid It	£3.95
BP311	An Introduction to Scanners and Scanning	£4.95
BP312	An Introduction to Microwaves	£3.95
BP313	A Concise Introduction to Sage	£3.95
BP314	A Concise Introduction to Quattro Pro	£4.95
BP315	An Introduction to the Electromagnetic Wave	£4.95
BP316	Practical Electronic Design Data	£4.95
BP317	Practical Electronic Timing	£4.95
BP318	A Concise User's Guide to MS-DOS 5	£4.95
BP319	Making MS-DOS Work for You	£4.95
BP320	Electronic Projects for Your PC	£4.95
BP321	Circuit Source – Book 1	£4.95
BP322	Circuit Source – Book 2	£4.95
BP323	How to Choose a Small Business Computer System	£4.95
BP324	The Art of Soldering	£3.95
BP325	A Concise Users Guide to Windows3.1	£4.95
BP326	The Electronics of Satellite Communications	£4.95
BP327	MS-DOS One Step at a Time	£4.95
BP328	More Advance Uses of Sage	£4.95